T0254558

SpringerBriefs in Petroleum Geoscience & Engineering

Series editors

Dorrik Stow, Heriot-Watt University, Edinburgh, UK
Mark Bentley, AGR TRACS Intern. Ltd, Aberdeen, UK
Jebraeel Gholinezhad, University of Portsmouth, Portsmouth, UK
Lateef Akanji, University of Aberdeen, Aberdeen, UK
Khalik Mohamad Sabil, Heriot-Watt University, Putrajaya, Malaysia
Susan Agar, ARAMCO, Houston, USA
Kenichi Soga, University of California, Berkeley, CA, USA
A. A. Sulaimon, Universiti Teknologi PETRONAS, Seri Iskandar, Malaysia

The SpringerBriefs series in Petroleum Geoscience & Engineering promotes and expedites the dissemination of substantive new research results, state-of-the-art subject reviews and tutorial overviews in the field of petroleum exploration, petroleum engineering and production technology. The subject focus is on upstream exploration and production, subsurface geoscience and engineering. These concise summaries (50–125 pages) will include cutting-edge research, analytical methods, advanced modelling techniques and practical applications. Coverage will extend to all theoretical and applied aspects of the field, including traditional drilling, shale-gas fracking, deepwater sedimentology, seismic exploration, pore-flow modelling and petroleum economics. Topics include but are not limited to:

- Petroleum Geology & Geophysics
- Exploration: Conventional and Unconventional
- Seismic Interpretation
- Formation Evaluation (well logging)
- Drilling and Completion
- Hydraulic Fracturing
- Geomechanics
- Reservoir Simulation and Modelling
- Flow in Porous Media: from nano- to field-scale
- Reservoir Engineering
- Production Engineering
- Well Engineering; Design, Decommissioning and Abandonment
- Petroleum Systems; Instrumentation and Control
- Flow Assurance, Mineral Scale & Hydrates
- Reservoir and Well Intervention
- Reservoir Stimulation
- Oilfield Chemistry
- Risk and Uncertainty
- Petroleum Economics and Energy Policy

Contributions to the series can be made by submitting a proposal to the responsible Springer contact, Charlotte Cross at charlotte.cross@springer.com or the Academic Series Editor, Prof. Dorrik Stow at dorrik.stow@pet.hw.ac.uk.

More information about this series at http://www.springer.com/series/15391

Marcelo Anunciação Jaculli
José Ricardo Pelaquim Mendes

Dynamic Buckling
of Columns Inside Oil Wells

 Springer

Marcelo Anunciação Jaculli
School of Mechanical Engineering
University of Campinas
Campinas, São Paulo
Brazil

José Ricardo Pelaquim Mendes
School of Mechanical Engineering
University of Campinas
Campinas, São Paulo
Brazil

ISSN 2509-3126   ISSN 2509-3134 (electronic)
SpringerBriefs in Petroleum Geoscience & Engineering
ISBN 978-3-319-91207-3   ISBN 978-3-319-91208-0 (eBook)
https://doi.org/10.1007/978-3-319-91208-0

Library of Congress Control Number: 2018940634

Printed on acid-free paper

This Springer imprint is published by the registered company Springer International Publishing AG part of Springer Nature
The registered company address is: Gewerbestrasse 11, 6330 Cham, Switzerland

To Angela, Marcos and Gisele, three pillars that sustain me and give me strength to always push forward. Love you very much.
Marcelo Anunciação Jaculli

To Nina, Davi and Luiza.
José Ricardo Pelaquim Mendes

Preface

Drilling and completion operations are fundamental to the oil industry. Through drilling, we connect the surface with oil reserves deep down underground by making holes; meanwhile, through completion, we prepare these holes to produce the oil in a secure and efficient way. There is a great number of equipment involved in these operations, and understanding how this equipment behaves when in use is mandatory to ensure safety.

The objective of this book is to describe the dynamic buckling behavior of columns inside directional wells through mathematical models. Such models can work together with current models in the literature, since the formers are dynamic, while the latter are static. In addition, by using this dynamic model, differences observed in practice regarding friction during tripping in and tripping out operations can be explained. Therefore, this book is useful for students and researchers studying topics related to column vibrations applied to the oil industry, especially for topics related to directional wells.

To understand better the proposed model and its implications, we divided this book into four chapters. In Chap. 1, we give a context regarding the usage of directional wells in the oil industry, highlighting the importance of studying operations related to them. In Chap. 2, we present the problem fundamentals, by giving an extensive literature review regarding the subject and by presenting basic concepts on column vibrations and column buckling. In Chap. 3, we present the models themselves, divided into four steps; each step represents a gradual progress in complexity starting from a base model, thus explaining the role of each hypothesis on the final model. Finally, in Chap. 4, we show the methodology and apply it on a simple study case for a directional well scenario, with proper results presented and discussed. We also do our final remarks, further commenting the results for the proposed problem and giving advice for future works regarding the subject.

We would like to acknowledge the support given by the University of Campinas and by CAPES during the development of this research and, subsequently, this book.

Campinas, Brazil   Marcelo Anunciação Jaculli
                                                             José Ricardo Pelaquim Mendes

Contents

Abbreviations and Symbols

ANP	Agência Nacional do Petróleo, Gás Natural e Biocombustíveis (Brazilian National Petroleum, Natural Gas and Biofuels Agency)
A	Cross sectional area (m^2)
$\hat{a}$	Acceleration vector (m/s^2)
$\hat{b}$	Binormal unitary vector (–)
BHA	Bottom Hole Assembly
C_f	Friction coefficient (–)
DPS	Dynamic Positioning System
dx	Infinitesimal element of length (m)
E	Young's modulus (N/m^2)
F	Internal force (N)
$\vec{F}$	Internal force vector (N)
$\vec{F}_f$	Total friction force (N)
$\vec{F}_{f1}$	Friction force component in the axial direction (N)
$\vec{F}_{f2}$	Friction force component in the tangential direction (N)
F_x	Internal force component in the $\hat{i}$ direction (N)
F_r	Internal force component in the $\hat{p}$ direction (N)
F_y	Internal force component in the $\hat{j}$ direction (N)
F_z	Internal force component in the $\hat{k}$ direction (N)
F_θ	Internal force component in the $\hat{q}$ direction (N)
$\vec{f}$	External forces per unit of length vector (N/m)
f	Total dynamic friction coefficient (–)
f_1	Dynamic friction coefficient in the axial direction (–)
f_2	Dynamic friction coefficient in the tangential direction (–)
f(x)	Function of a single variable (–)
g	Gravitational acceleration (m/s^2)
$\vec{H}_0$	Angular momentum vector ($kg*m^2/s$)
h	Interval for discretization (–)

I	Area moment of inertia (m^4)
I_p	Mass moment of inertia per unit of length (kg*m)
$\hat{i}$	Axial direction vector (direction vector in x axis) (–)
i	Space index for discretization (–)
$\hat{j}$	Direction vector in y axis (–)
j	Time index for discretization (–)
$\hat{k}$	Direction vector in z axis (–)
k	Vector norm of $\partial\vec{\tau}/\partial x$ (–)
KOP	Kickoff Point
k_r	Component of $\vec{\tau}$ in the $\hat{p}$ direction (–)
k_θ	Component of $\vec{\tau}$ in the $\hat{q}$ direction (–)
L	Column length (m)
M	Internal moment (N*m)
$\vec{M}$	Internal moment vector (N*m)
M_r	Internal moment component in the $\hat{p}$direction (N*m)
M_θ	Internal moment component in the $\hat{q}$direction (N*m)
m_p	Mass per unit of length (kg/m)
$\vec{N}$	Normal force vector (N)
N	Normal contact force per unit of length (N/m)
$\vec{n}$	Normal unitary vector (–)
$\vec{P}$	Linear momentum vector (kg*m/s)
$\hat{p}$	Normal direction vector (–)
$\vec{q}_p$	Weight force vector (N)
$\hat{q}$	Tangential direction vector (–)
$\vec{r}$	Position vector (m)
r	Clearance between the column and the well (m)
RAO	Response Amplitude Operator
r_c	Well radius (m)
T	Total time interval (s)
t	Time variable (s)
U_h	Heave amplitude (m)
u(x,t)	Axial displacement function (m)
u_a	Axial displacement due to axial tension/compression (m)
u_b	Axial displacement due to bending (m)
u_x	Total axial displacement (m)
V	Internal shear force (N)
$\vec{V}$	Velocity vector (m/s)
v_1	Velocity in the axial direction (m/s)
v_2	Velocity in the tangential direction (m/s)
w	Lateral displacement (m)
WOB	Weight-on-bit (–)

x	Space coordinate in the axial direction (m)
α	Well inclination angle (rad)
Δt	Time interval for discretization (s)
Δx	Space interval for discretization (m)
ε	Axial strain (–)
θ	Angular displacement (rad)
ρ	Material density (kg/m^3)
σ	Axial stress, (N/m^2)
$\vec{\tau}$	Position unitary vector (–)
$\vec{\Omega}$	Angular velocity vector (rad/s)
ω	Column rotational angular frequency (rad/s)
ω_h	Heave angular frequency (rad/s)
ω_r	Component of $\vec{\Omega}$ (rad/s)
ω_θ	Component of $\vec{\Omega}$ (rad/s)

Chapter 1
Introduction

Since the technology of directional drilling was improved in the USA in the 70s, directional wells became the reality of the oil industry. Directional and horizontal wells have innumerous advantages over vertical ones, such as increasing the well productivity by increasing the area in contact with the reservoir; drilling multiple development wells from a single platform, thus lowering costs; reaching hard objectives, such as formations below urban and environmental protected areas; sidetracking, an operation in which the well is deviated from its original trajectory in order to avoid anything restricting the path such as "fishes"; exploring fractured reservoirs; drilling relief wells for controlling blowouts; and multilateral wells, which are wells with several "legs" produced from different zones. Therefore, directional wells are extremely important for the oil industry and understanding the behavior of all equipment during operations of directional drilling and completion becomes vital to ensure safety.

Here in this book, to show the importance of directional wells, we will focus on well data gathered from Brazilian basins, which can be found easily at ANP's (the Brazilian National Petroleum, Natural Gas and Biofuels Agency) website. Figure 1.1 shows the number of directional and horizontal wells drilled on Brazilian basins for selected year periods. The 2011–2017 period is marked in red as a reminder that this period is not a full decade as the others on the graph (in fact, a simple projection indicates that this period will surpass the previous decade in number of wells drilled).

As can be seen from Fig. 1.1, few directional and horizontal wells were drilled before the 70s; this was due to the necessary technology still being developed, but also due to Brazilian oil exploration being focused on onshore fields, which could be done with vertical wells only. Serious drilling of directional and horizontal wells began in the 70s, increasing a lot during the 80s and 90s, and then exploding in the 2000s, thanks to technological advancements, but also due to increased investments on the Brazilian oil sector.

An interesting way to see how directional drilling expanded is using some filters for the well data. Figure 1.2 shows the distribution by well type under several criteria: all wells in ANP's database included, which amounts to 29575 wells; exploitation

© The Author(s) 2018
M. A. Jaculli and J. R. P. Mendes, *Dynamic Buckling of Columns Inside Oil Wells*,
SpringerBriefs in Petroleum Geoscience & Engineering,
https://doi.org/10.1007/978-3-319-91208-0_1

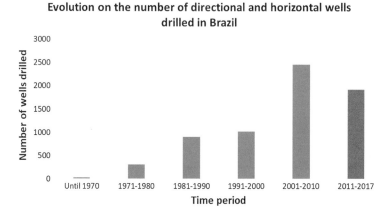

Fig. 1.1 Number of directional and horizontal well drilled during each decade. Data obtained from ANP (2018)

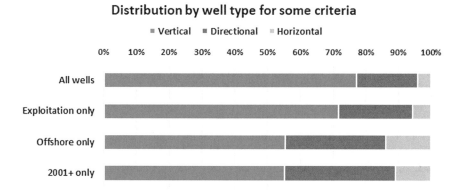

Fig. 1.2 Distribution by well type according to filters: all wells, exploitation wells only, offshore wells only and 2001 + wells only. Data obtained from ANP (2018)

wells only, i.e., only wells drilled during the development of the field for either production or injection, which amounts to 21241 wells; offshore wells only, which amounts to 6649 wells; and finally, only wells drilled during and after 2001, which amounts to 9793 wells.

As seen in Fig. 1.2, after applying filters, directional and horizontal wells become more relevant compared to vertical wells; this is a direct consequence of the chosen filters. First, vertical wells are often employed during the exploration phase, since you do not have much information about the field and they are cheaper to drill; meanwhile, directional and horizontal wells are more employed during the exploitation phase, since they possess the aforementioned advantages for oil production. Second, vertical wells are usually drilled more on onshore basins—whose reservoirs are not very deep—while directional and horizontal wells are drilled on offshore

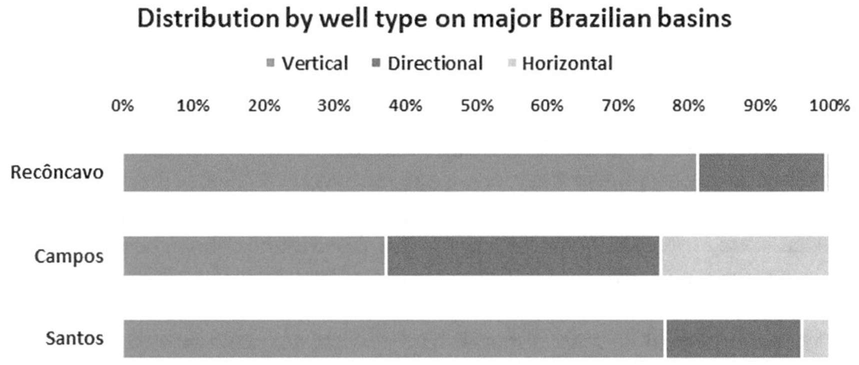

Distribution by well type on major Brazilian basins

■ Vertical ■ Directional ■ Horizontal

Fig. 1.3 Distribution by well type for three Brazilian basins: Recôncavo basin (mainly onshore), Campos basin (offshore, mainly post-salt), and Santos basin (offshore, mainly pre-salt). Data obtained from ANP (2018)

basins—which are usually deep and full of obstacles. Last, vertical wells were more predominant in the past, when the directional drilling technology was still being developed; nowadays, with the technology mastered, directional and horizontal wells are preferred over vertical ones due to intrinsic advantages.

Another interesting aspect is the distribution according to the basin. Figure 1.3 shows the distribution by well type for three Brazilian basins: Recôncavo, which is mainly onshore; Campos, which is offshore and the major basin in Brazil; and Santos, which contains the Brazilian pre-salt and is relatively new.

As seen in Fig. 1.3, the onshore basin (Recôncavo) has a much higher percentage of vertical wells than the offshore basin (Campos), further supporting the fact that directional drilling is more employed on offshore basins as mentioned previously. The Santos basin is a special case: despite being offshore, it is still a relatively new basin, thus possessing a high vertical wells percentage due to the early exploration activities, but also because vertical wells are still preferred for drilling on salt formations.

Considering all the data showing the relevance of directional wells in the oil industry, this book thus focuses on the dynamic behavior of columns used on completion operations inside directional wells. During completion operations, it was observed that the friction on a column—measured indirectly through the hook load—was different during operations of tripping in and out of the well. Several completion operations involve the use of a column inside another column, such as lowering a tubing string inside a cased hole; a coiled tubing string inside a tubing string; or a sand screen using a work string inside an open hole. This problem originally appeared when measurements of hook load during tripping in and out—which were taken since tubing auxiliary lines were failing—indicated that the friction force would be different in both cases. This was unexpected because none of the available models and software could explain this effect or even consider that the column can have different friction forces. Since this problem did not happen with drill strings—which are stiffer than a tubing and a coiled tubing—the cause would probably be associated with buckling. During tripping in, the tubing is being lowered inside the well using its own weight, and compressive forces can act on it due to contact with the well-bore; meanwhile, during tripping out, the tubing is being pulled from the well, thus it is subjected to a tension force instead. Therefore, since forces occur in different directions during tripping in and tripping out, the hypothesis here is that the tubing is suffering buckling during its tripping in due to compression, while the tubing is not buckled during tripping out due to tension. The fact that the column buckles in only one scenario could explain the difference between the friction forces. Another thing to consider is that the problem may be related directly to the dynamic response of the system instead of its static response. The problem could not be explained by the most common commercial software available in the market—considering that they are based on analyzing the static behavior of the column—thus leading us authors to believe that the cause is dynamical. Therefore, the aforementioned columns vibrating during such operations while being constrained by the well or another column are the object of study of this book.

1.1   Summary

This chapter:

- Discussed the importance and applications of directional wells in the oil industry;
- Showed that directional wells become more relevant in offshore environments and in field development stages.

Reference

ANP. Tabela de Poços (Wells Table) (2018). In: Acesso aos Dados Técnicos (Access to Technical Data). http://www.anp.gov.br/wwwanp/exploracao-e-producao-de-oleo-e-gas/dados-tecnicos/acesso-aos-dados-tecnicos (in Portuguese). Accessed on 21 Feb 2018

Chapter 2
Fundamentals on Column Dynamics

In this chapter, a literature review regarding the subject is made. First, a historical background about the problem of vibrations in columns is presented, and then a brief review regarding directional drilling is also made. The buckling problem itself is presented next, by showing first the literature regarding the static approach, and then recent works for the dynamic approach.

2.1 Literature Review

2.1.1 Historical Background

The exploration of offshore fields and the construction of deep and directional wells brought the necessity to understand the behavior of columns under such conditions—whether they are risers, drilling strings, or tubing strings. Several works regarding the subject were published starting from the 60s, and remains the focus of intense studies up until now. Bailey and Finnie (1960) and Finnie and Bailey (1960) are perhaps the pioneers on the subject of the behavior of columns. Since then, there were many other works about columns vibrations, characterizing its three vibration modes—axial, torsional, and lateral—as well as the coupling between these three modes. On axial vibrations, the ones that stand out are Chung and Whitney (1981), Sparks et al. (1982), and Niedzwecki and Thampi (1988), whereas, on lateral vibration, there are Park et al. (2002) and Sparks (2002). It is evident that there are several other works exploring the topic. However, only a few hypotheses and/or boundary conditions change, always keeping the essence of the original problem—for axial and torsional vibrations, the wave equation first proposed by Jean d'Alembert in 1746; for lateral vibrations, the beam models deducted by Leonhard Euler and Daniel Bernoulli around 1750 and by Stephen Timoshenko in 1921. Han and Benaroya (2002) studied the behavior of columns using the Euler–Bernoulli and Timoshenko

© The Author(s) 2018
M. A. Jaculli and J. R. P. Mendes, *Dynamic Buckling of Columns Inside Oil Wells*,
SpringerBriefs in Petroleum Geoscience & Engineering,
https://doi.org/10.1007/978-3-319-91208-0_2

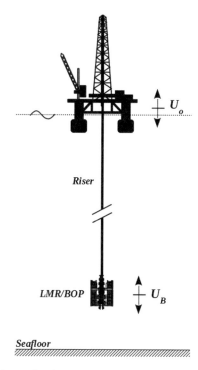

Fig. 2.1 Example of a column vibrating on an offshore environment

beam models, whereas, more recently, Chin (2014) studied the effect caused by the coupling of the three vibration modes; both works also presented numerical solutions for the motion equations of their respective problems. Chakrabarti (2003) presents the necessary modeling to study the dynamic behavior of offshore structures, such as vessels and floating platforms; such analysis is needed for the study of offshore wells, since the motion of such structures is transmitted to the column, thus making the offshore problem essentially different from an onshore one—where there is no such kind of motion. An example of a problem associated with column vibrations can be seen in Fig. 2.1. The figure exemplifies a common operation during drilling wells: the drilling rig—represented by the floating vessel—lowers a BOP—represented by the lumped mass at the bottom—using a riser—represented by the column. As the floating vessel vibrates with amplitude U_0, caused by the ocean waves, the motion is transmitted through the column to the lumped mass below, which vibrates with a different amplitude U_B.

Despite the progress on studying vibrations of such systems—a problem which is intrinsically dynamic—there were still phenomena associated with the static problem, such as column buckling. Lubinski et al. (1962) published one of the first works on the subject. The initial concern was only for vertical wells; in such cases, there was the possibility that the tubing string would buckle due to loadings caused by

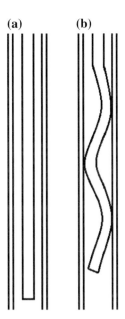

Fig. 2.2 Example of (**a**) non-buckled and (**b**) buckled tubing string inside a well

temperature and fluid pressure. Since the column was virtually fixed on its lower end due to the packer, such loadings could cause compression, and consequently buckling. In the case that buckling would occur, the column would suffer changes on its length, which would be reduced; therefore, the authors focused on how to estimate such changes, calculating the so-called effective length. Figure 2.2 exemplifies this effect, showing the length reduction caused by buckling.

As the construction of wells became more complex, with the beginning of the practice of building directional wells, the solutions for vertical wells were no longer enough to describe the column behavior. Paslay and Bogy (1964) and, later, Dawson and Paslay (1984) noted that a model which considered the effects of the well inclination on buckling was needed. For such, the authors deducted a formula to calculate the critical buckling force—the maximum compressive force that the column could resist without suffering buckling—considering the well inclination. This criterion for the critical force is still widely accepted for solving problems associated to buckling, being used on commercial software. Works after Dawson and Paslay (1984) tried to improve the criterion by including, for example, the influence of friction.

Last, more recently, Gao and Miska (2010a) published a work analyzing the dynamic behavior of a column in an already buckled configuration. Through the dynamic analysis, it is possible to compare the column behavior under a buckled condition with its non-buckled condition. These differences can explain phenomena observed in practice and/or experimentally, such as the differences in the friction force during the column tripping in and during its tripping out.

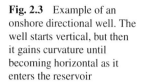

Fig. 2.3 Example of an onshore directional well. The well starts vertical, but then it gains curvature until becoming horizontal as it enters the reservoir

Evidently, the interest in both the vibration of continuous systems and the buckling of columns is not exclusive to the oil exploration activities. Timoshenko (1937) had already presented the equations for the axial vibration of bars, torsional vibration of shafts, and lateral vibrations of beams long before any other work mentioned here; in 1757, Euler had already deducted the Euler's critical load, the maximum load that a column could support without buckling. Rao (2007) gathers the motion equations for the most common continuous systems in engineering applications, as well as the analytical methods of Newton and Lagrange, used to deduct the equations.

2.1.2 Directional Drilling

Since this book is based on directional wells, a short review regarding directional drilling is in place. A classical book regarding drilling is the one from Bourgoyne et al. (1986), while Rocha et al. (2006) focused specifically on directional drilling.

In Bourgoyne et al. (1986), several aspects of drilling are discussed, such as required equipment, drilling fluids, cementing, and drilling hydraulics. The interesting part of this book is chapter 8, in which directional drilling is discussed. The authors provide insight on the technology, such as applications, special equipment, and trajectory planning and control. Meanwhile, in Rocha et al. (2006), a much more in-depth discussion is made, especially regarding the practices adopted in Brazil. The book provides examples of directional wells trajectories, which will be used in this book. Figure 2.3 illustrates an onshore directional well.

2.1.3 Column Buckling—Static Approach

As previously mentioned, the concern of authors regarding the buckling phenomenon was mostly with respect to the static problem, not the dynamic one. Therefore, until today, the majority of published works—as well as the commercial software developed specially for this kind of problem—worried only about the static approach, with more or less the same goal: to estimate which compressive force will cause buckling—i.e., the critical buckling load—and to find the new column length after buckling—i.e., the effective length. What distinguishes the innumerous works in the literature are the possible nuances in the model: different boundary conditions, different column configurations, the effect of dry friction, the effect of well inclination. Among these innumerous works, the ones that stand out are the pioneers and still widely referenced today: (Lubinski et al. 1962; Paslay and Bogy 1964; and Dawson and Paslay 1984). It is interesting to note as well the work done by Mitchell (2008) who did a summary of the state of the art column buckling, by narrating the history of publications on the subject and enunciating the challenges that remain to be analyzed to understand it even better.

Lubinski et al. (1962) made an analysis on the effect of internal and external pressures and temperature on the static behavior of a tubing string. According to them, the column might suffer helical buckling when the packer used to settle it can seal its motion completely, but also when the motion is permitted. In case buckling occurs, the original length of the column will reduce. Therefore, the authors propose models to calculate the column length reduction and explored practical cases for each one of the possible scenarios: packer without any column motion restrictions, packer with partial column motion restriction, and packer with complete column motion restriction and permanent corkscrewing—phenomenon in which the column suffers plastic strain and retains the helical configuration even under tension. These calculations aim to mitigate the buckling effects during the tubing string installation and/or operation, thus predicting, for example, which maximum compressive force can be applied to avoid buckling. The authors also remark that buckling can cause operational problems even when it does not cause tubing failure. If the tubing buckles, the passage of tools using wireline may become impossible.

Paslay and Bogy (1964), using energy methods, do an extensive analysis regarding the stability of a bar subjected to tension loads and confined inside an inclined cylinder. Considering the hypothesis that the bar always remains in contact with the internal surface of the cylinder, the authors conclude that the bar will always be stable, as long as there are no restrictions for its rotational motion. They also concluded that for a bar of small diameter—compared to the diameter of the external cylinder—the confinement effect becomes negligible, and thus the critical buckling load, in this case, reduces itself to the traditional Euler's critical load. These results were fundamental for subsequent works to elaborate more robust criteria regarding the critical buckling load.

Dawson and Paslay (1984) suggest corrections for the work of Paslay and Bogy (1964), aiming to consider the effect of floatability of the drill string. As observed by the authors, the critical load for a column calculated by Paslay and Bogy (1964) for slant wells results in a higher value than for the same column inside a vertical well; therefore, the drill string is more resistant to buckling in directional wells than on vertical ones. Since the column is more resistant in that scenario, there is the possibility of tripping in under compression on the slant segment without the risk of buckling. In addition, with a lesser risk of buckling, the BHA (bottom hole assembly) weight can be reduced, which in turn will reduce the torque and drag during operations as well. Even so, there is still the possibility that the column buckles under an excessive compressive load; therefore, the authors develop a criterion to avoid buckling, taking into account the weight-on-bit (WOB) and the column wet weight. By controlling these two variables, it is possible to avoid that the compressive load surpasses the critical buckling load. Last, the authors remark that, as the time passes, the mechanical properties of the joints degrade, becoming less rigid, lighter, and consequently, more susceptible to buckling, thus influencing on the results obtained through usage of the buckling models. In addition, the analyses are valid only for slant segments of well—which have constant inclination—because drastic changes on the inclination can compromise the column resistance to buckling by reducing the critical buckling load.

Based on these classical works, there is a vast list of other works, which gives contributions starting from these initial models. Following up, only a few of these works are presented, the ones that we consider relevant in building the knowledge on the subject. They are organized in chronological order but also grouped based on the kind of contribution given.

Mitchell (1986) makes a simplified analysis aiming to consider the effect of the dry friction force on the critical buckling load. The author concludes that the friction force reduces the compressive force acting on the tubing string, thus attenuating the effects of buckling. If the buckling effect is attenuated, the column original length will not reduce as much as predicted on previous works, which in turn gives more freedom when designing packers—a problem which had been already identified by Lubinski et al. (1962). However, the author also concludes that his model still needs improvements. He made such improvements in Mitchell (1996b), in which the model also considers the load history on the column—for the case in which the column is loaded once, the load is removed, and then a new load is applied—thus being able to pinpoint the direction of the friction force during the second loading. Later, Mitchell (2007), knowing that the column could either slip on the wellbore or roll without slipping, modifies Dawson and Paslay's (1984) criterion to consider such effects; the difference between each criterion is the presence of a term respective to the torsional rigidity. Following Mitchell's (2007) footsteps, Gao and Miska (2009) recognize that the friction force possesses components in more than one direction; there is lateral friction due to the angular motion of the column inside the well, as well as axial friction due to the axial strain of the column.

Chen et al. (1990) are perhaps the first ones to recognize and distinguish the existence of two different buckling modes: sinusoidal—also commonly called lateral—and helical. Figure 2.4 shows the two modes of buckling. The authors then establish criteria that would separate the two kinds of buckling and could recognize

Fig. 2.4 Example of **a** sinusoidal and **b** helical buckling

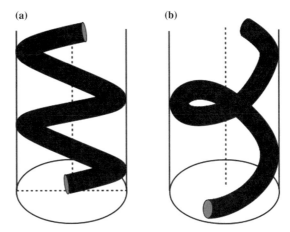

which one would happen first. They conclude that, first, sinusoidal buckling would happen when the critical buckling load is reached and only then helical buckling could happen, which requires an even larger load than the critical load. The authors also find out the existence of another critical value even higher, in which a phenomenon called lock-up would occur: the column would lock inside the well in its helical configuration and would not be able to move on the axial direction any longer, even under tension.

Saliés (1994) makes an experimental study to measure the critical buckling load. A schematic of his experiment can be seen on Fig. 2.5. The author varies several parameters to observe the effect of each one of them on the final result: different pipe diameters and thickness, different materials for different dry friction coefficients, different well inclinations ranging from vertical until horizontal. The author then compared the obtained results with the existing models in the literature, such as Lubinski et al. (1962), Dawson and Paslay (1984) and Chen et al. (1990). The experimental results are satisfactory, with good congruence when compared to the values calculated from the models. The author also concludes that the friction, aside from increasing the critical buckling load as already observed previously by Mitchell (1986), also creates a hysteresis effect during loading and unloading of the pipe. Last, he also concludes that the tendency is for the column to suffer helical buckling, with sinusoidal buckling happening only in its first mode before it moves to helical.

He and Kyllingstad (1995) improve the model from Dawson and Paslay (1984) to consider the well curvature. Up until then, Dawson and Paslay's (1984) model only considered that the column was on an inclined well segment with constant curvature; the consequence of such model was that the critical buckling load calculated from it was still too conservative when compared with measured data. For this very reason, the authors consider the effect of the well curvature, which increases the critical buckling load and thus is less conservative. Later, Mitchell (1999) reformulated the criterion for the critical buckling load initially proposed by Dawson and Paslay (1984) by taking into account the work from He and Kyllingstad (1995).

Fig. 2.5 Schematic of an
experiment for measuring
the critical buckling load

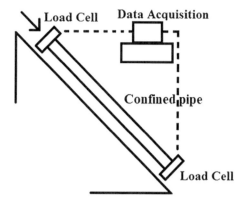

Using the Euler–Bernoulli slender beam model, Mitchell (1988) proposes an equilibrium equation to calculate the static displacements of a column already under buckling inside a vertical well. After that, in Mitchell (1996a) and Mitchell (1997), the author improves the model to consider also directional wells. Last, Mitchell (2002) seeks analytical solutions for the presented equilibrium equations, especially for the cases of vertical and horizontal wells. Starting from Mitchell's (1988) model, Gao and Miska (2009) study the effects of the boundary conditions and friction force on the static configuration of the column after buckling occurred. They concluded that for slender pipes, the effect of the boundary conditions can be neglected without affecting the result, whereas the effect of the friction force becomes even more relevant since the critical buckling load is increased—something that was already mentioned by Mitchell (1986). Similar to Paslay and Bogy (1964), they also conclude that for non-slender pipes the effect of the wellbore could be neglected. These results are expanded in Gao and Miska (Gao and Miska 2010b).

Miska and Cunha (1995) performed an extensive analysis regarding the critical buckling load, considering six different combinations: whether the column had weight or not, combined with either pure axial loading, pure torsional loading or both loads. The authors note that the torque reduces the critical buckling load, besides also reducing the helix pitch during helical buckling. Such effects are more noticeable in wells with smaller inclinations or in columns that are more flexible. These results are expanded later in Qiu et al. (1998, 1999), where the authors improve the model from Miska and Cunha (1995) for the 3D case, besides analyzing the influence of the column initial configuration on the buckling phenomenon. Wicks et al. (2007) propose a critical buckling load criterion for long cylinders by taking into account the effects of compression and torsion, while also concluding that more studies are required to include the gravity and friction effects properly as well.

Mitchell (2008) makes a summary of the state of the art column buckling problem. The author presents the criteria for the critical buckling load developed by Dawson and Paslay (1984), Chen et al. (1990), and He and Kyllingstad (1995), with respect to the two possible buckling configurations: sinusoidal and helical. It also shows the corrections to consider the friction effect, obtained by Mitchell (2007). Last, the

Fig. 2.6 Example of a tapered-string problem

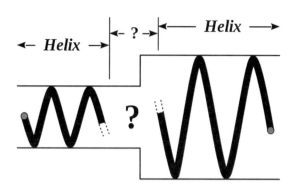

author presents the equilibrium equations for the column after buckling, initially obtained by Lubinski et al. (1962) and expanded by Miska and Cunha (1995) and Mitchell (2002). As for remaining challenges, the author mentions the modeling of columns with segments of different properties known as tapered strings, shown in Fig. 2.6, which bring uncertainties to the problem due to the change of diameter; the effect of the boundary conditions on directional wells; and fully understanding the role of the friction force on the problem. Despite the fact that papers such as Mitchell (1986) approached the effect of the friction force, the author judges that this effect is not completely described and understood.

Recent works have focused on comparing results obtained from literature models with experimentally measured data, such as in Arslan et al. (2014); or improving even further the models for the static configuration after buckling, as in Huang et al. (2015a, b).

2.1.4 Column Buckling—Dynamic Approach

Different from the static problem, the dynamic problem associated to column buckling has received little attention from authors. However, there are two plausible explanations for this fact. First, the dynamic analysis has little contribution in creating criteria to evaluate if the column will buckle or not. This happens because the dynamic analyses already consider that the column will buckle regardless, similar to what was done by Gao and Miska (2009) for the column configuration after buckling on the static case. Second, the motion equations describing the column become complex due to the coupling that appears between axial and angular displacements, resulting on a system of nonlinear partial differential equations. Analytical solutions become impossible—unless several simplifications are made—thus numerical methods being the only possible path to follow.

Gao and Miska (2010a) is the most relevant work regarding the dynamic approach. There, the authors deduct a dynamic model to describe the vibration of a column under an already buckled condition. Such model results in a system of partial differential

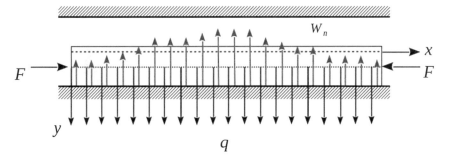

Fig. 2.7 Example of a column inside a horizontal segment of well. W_n is the distributed normal force, q is the column distributed weight and F is a compressive force

equations relating axial displacement, angular displacements, axial internal force, and normal contact force between the column and the well. After several simplifications, the authors are able to find an analytical solution and analyze the phenomena that occur during the vibration of the buckled column. They observe that depending on the amplitude of vibration, the column might have two different behaviors, called the first and second modes of snaking motion. On the first mode, the column vibrates only in contact with half of the well; in other words, starting from the equilibrium position at the lowest point of the well, it can vibrate and reach the highest point of the well only in contact with one of the two sides. Meanwhile, on the second mode, the column is free to vibrate in contact with any point of the well. The authors conclude that the model still needs improvements, since it neglects the dry friction force, which is most likely relevant to the phenomenon. In Sun et al. (2014), the authors expand Gao and Miska's (2010a) work, finding approximated analytical solutions and comparing with numerical solutions, reaching good results. Figure 2.7 shows a schematic of the problem proposed by Sun et al. (2014).

Despite a robust model already existing to explain the dynamic problem associated to column buckling, there is still a lot to be done. As explained by Gao and Miska (2010a), the two biggest simplifications of their model are the friction force being neglected and the analysis is valid only for a horizontal well. The goal of this book is exactly to push forward on these two hypotheses while also exploring the implications of what was deducted already by them.

2.2 Column Vibration Fundamentals

Before the problem of column buckling is presented, a quick review on the topic of column vibration is in place. In order to understand the fundamentals behind column buckling, here the axial vibration of bars and the lateral vibration of beams are presented.

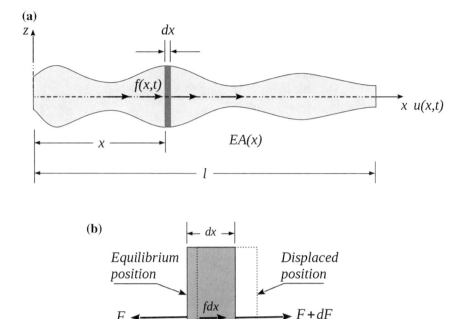

Fig. 2.8 (**a**) Bar with variable cross section. (**b**) Infinitesimal element of bar, subjected to internal axial forces

2.2.1 Axial Vibration of Bars

Given a bar, let L be the bar length, A(x) the variable cross-section, E the Young's modulus, and f(x) an applied external force per unit of length, as seen in Fig. 2.8a. After obtaining an infinitesimal element dx by cutting the bar along its cross-section, as seen in Fig. 2.8b, two internal axial forces F and F + dF will appear.

Let u be the bar axial displacement. Applying Newton's Second Law on the infinitesimal element dx, Eq. 2.1 is obtained:

$$F + dF - F + f(x, t)dx = \rho A(x)dx\frac{\partial^2 u}{\partial t^2} \qquad (2.1)$$

where ρ is the bar material density. The internal axial force F is directly related to the axial stress and with Hooke's Law through Eq. 2.2:

$$F = \sigma A(x) = E\varepsilon A(x) = EA(x)\frac{\partial u}{\partial x} \tag{2.2}$$

where σ is the axial stress and ε is the axial strain. Knowing that the differential of axial force can be represented as in Eq. 2.3:

$$dF = \frac{\partial F}{\partial x}dx \tag{2.3}$$

Substituting Eq. 2.2 into Eq. 2.3, then on Eq. 2.1 and simplifying:

$$\rho A(x)\frac{\partial^2 u}{\partial t^2} - \frac{\partial}{\partial x}\left(EA(x)\frac{\partial u}{\partial x}\right) = f(x,t) \tag{2.4}$$

For a bar of constant cross-section $A(x)$, Eq. 2.4 reduces to:

$$\rho A\frac{\partial^2 u}{\partial t^2} - EA\frac{\partial^2 u}{\partial x^2} = f(x,t) \tag{2.5}$$

Equation 2.5 is the basic equation to study axial vibrations of bars, one of the motions that will compose the column buckling phenomenon with the lateral vibration of beams, presented in the following section.

2.2.2   Lateral Vibration of Beams

Given a beam, let L be the beam length, $A(x)$ the variable cross-section, E the Young's modulus, $I(x)$ the second moment of inertia, and $f(x)$ an applied external force per unit of length, as seen in Fig. 2.9a. After obtaining an infinitesimal element dx by cutting the bar along its cross-section, as seen in Fig. 2.9b, two internal shear forces V and $V+dV$, as well as two bending moments M and $M+dM$, will appear.

Let w be the beam lateral displacement. Applying Newton's Second Law on the infinitesimal element dx, Eq. 2.6 is obtained:

$$V - (V + dV) + f(x,t)dx = \rho A(x)dx\frac{\partial^2 w}{\partial t^2} \tag{2.6}$$

where ρ is the beam material density. The summation of moments on point P from Fig. 2.9b will lead to:

$$M + dM - M - (V + dV)dx + f(x,t)dx\frac{dx}{2} = 0 \tag{2.7}$$

Knowing that the differential of shear force and bending moment can be written as:

$$dV = \frac{\partial V}{\partial x}dx \tag{2.8}$$

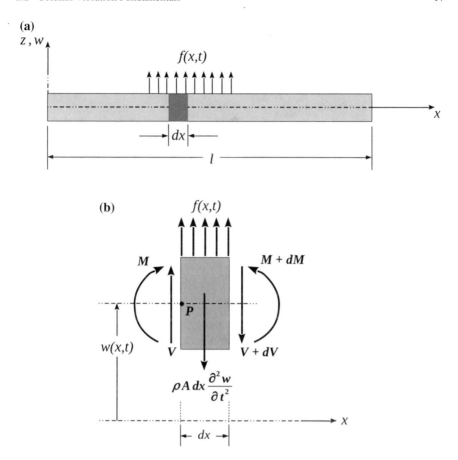

Fig. 2.9 (**a**) Beam with variable cross section. (**b**) Infinitesimal element of beam, subjected to internal shear forces and bending moments

$$dM = \frac{\partial M}{\partial x} dx \tag{2.9}$$

Substituting Eqs. 2.8–2.9 in Eqs. 2.6–2.7 and neglecting terms of order dx^2 will lead to:

$$-\frac{\partial V}{\partial x} + f(x, t) = \rho A(x) \frac{\partial^2 w}{\partial t^2} \tag{2.10}$$

$$\frac{\partial M}{\partial x} - V = 0 \tag{2.11}$$

Substituting Eq. 2.11 in Eq. 2.10:

$$\rho A(x)\frac{\partial^2 w}{\partial t^2} + \frac{\partial^2 M}{\partial x^2} = f(x, t) \tag{2.12}$$

According to the Euler–Bernoulli beam theory, the bending moment on a certain position x along the beam is given by:

$$M = EI(x)\frac{\partial^2 w}{\partial x^2} \tag{2.13}$$

Substituting Eq. 2.13 into Eq. 2.12:

$$\rho A(x)\frac{\partial^2 w}{\partial t^2} + \frac{\partial^2}{\partial x^2}\left(EI(x)\frac{\partial^2 w}{\partial x^2}\right) = f(x, t) \tag{2.14}$$

For a bar of constant cross-section A(x)—and consequently constant moment of inertia I(x)—Eq. 2.14 reduces to:

$$\rho A\frac{\partial^2 w}{\partial t^2} + EI\frac{\partial^4 w}{\partial x^4} = f(x, t) \tag{2.15}$$

Equation 2.15 is the basic equation to study lateral vibrations of beams, one of the motions which will compose the column buckling phenomenon with the axial vibration of bars of the previous section. Equations 2.5 and 2.15 are not used directly, since these equations are uncoupled from each other and the nature of the buckling problem couples these two effects. Aside from these axial and lateral motions, there is the dynamic buckling motion itself, which will be explained in the following section.

Fig. 2.10 Column inside a well scheme. The column is represented by the smaller circle, while the well is represented by the larger circle

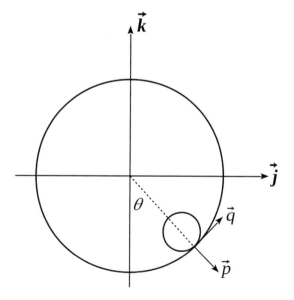

2.3   Column Buckling Fundamentals

In this section, we do a short introduction of the column buckling inside directional wells to familiarize the readers with the physics behind the subject. In Chap. 3, the problem and its associated models are developed further.

In a directional well, the column is free to vibrate in all three directions, aside from rotating around its own axis. To understand the physics, let us consider first a horizontal segment of well—this hypothesis is pushed further in subsequent models. If we consider that the column remains in contact with the well throughout its whole length and during the whole time, the number of variables is reduced from three—initially the displacements on the x, y, and z axis—to two—axial displacement along the well axis and angular displacement as defined by Fig. 2.10. To help with modeling, two unitary vectors $\hat{p}$ and $\hat{q}$ are defined—normal and tangential, respectively, to the contact point between the column and the wellbore, as seen in Fig. 2.10 as well. The angle θ is defined between the z axis of the well and the normal vector $\hat{p}$.

Figure 2.11 shows the column buckled configuration inside the well, from planes xz and yz, with plane yz containing the cross section and the x axis providing the

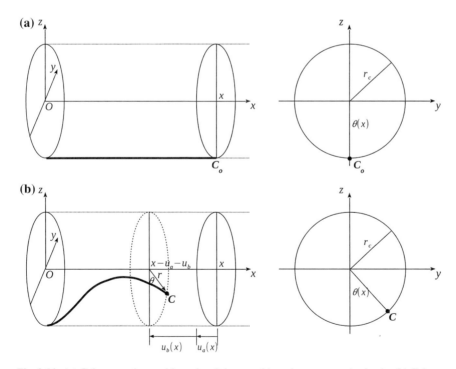

Fig. 2.11 (**a**) Column resting position when it is not subjected to compressive loads. (**b**) Column buckled position caused by compressive loads

position along the horizontal segment. Initially, the column is not subjected to any kind of load on the axial direction; consequently, it is not buckled and rests on the lowest portion of the well, such as in Fig. 2.11a. When a compressive force high enough to cause buckling is acting on the column, it suffers axial and angular displacements simultaneously. It is important to observe that the column final position is a consequence of both the axial contraction u_a and the contraction caused by bending u_b, as seen in Fig. 2.11b. Thus, to describe the dynamic buckling phenomenon, it is necessary to understand how the axial and angular displacements occur as a function of time.

One of the main ideas of the following models is that during tripping in, the column is being compressed and thus can suffer buckling, displacing itself angularly inside the well to form either a sinusoid or a helix, as seen in Fig. 2.11b; meanwhile, during tripping out, the column is under tension and thus there is no buckling, meaning that the column will remain in contact with the lowest portion of the well the whole time, as seen in Fig. 2.11a. These different behaviors will then lead to the differences in friction observed in practice, as mentioned in Chap. 1. The equations describing the phenomenon for all its models are shown on the following Chap. 3.

2.4 Summary

This chapter:

- Does a literature review on the subject, divided into topics:
  - Historical background on studies regarding column vibrations;
  - Directional drilling technology;
  - Column buckling, by considering a static approach;
  - Column buckling, by considering a dynamic approach.

References

Arslan M, Ozbayoglu EM, Miska SZ, Yu M, Takach N, Mitchell RF (2014) Buckling of buoyancy-assisted tubulars. SPE Drill Completion 29(4), 372–385, SPE 159747-PA. http://dx.doi.org/10.2118/159747-PA

Bailey JJ, Finnie I (1960) An analytical study of drill-string vibration. ASME J Eng Ind 82(2):122–127. https://doi.org/10.1115/1.3663017

Bourgoyne AT, Millheim KK, Chenevert ME, Young FS (1986) Applied Drilling Engineering. SPE Textbook Series, vol. 2. Society of Petroleum Engineers, Richardson, 502 pp

Chakrabarti SK (2003) Hydrodynamics of offshore structures. Wit Press/Computational Mechanics, Ashurst, 464 pp

Chen Y, Lin Y, Cheatham JB (1990) Tubing and casing buckling in horizontal wells. J Petroleum Technol 42(2): 140–191, SPE 19176-PA. http://dx.doi.org/10.2118/19176-PA

Chin WC (2014) Wave Propagation in Drilling, Well Logging and Reservoir Applications. Advances in Petroleum Engineering. Scrivener Publishing, Beverly, 456 pp

Chung JS, Whitney AK (1981) Dynamic vertical stretching oscillation of an 18000 ft ocean mining pipe. In: Offshore Technology Conference, 4–7 May, Houston, Texas. OTC 4092-MS. http://dx.doi.org/10.4043/4092-MS

Dawson R, Paslay PR (1984) Drillpipe buckling in inclined holes. J Petroleum Technol 36(10): 1734–1738, SPE 11167-PA. http://dx.doi.org/10.2118/11167-PA

Finnie I, Bailey JJ (1960) An experimental study of drill-string vibration. ASME J Eng Ind 82(2): 129–135. http://dx.doi.org/10.1115/1.3663020

Gao G, Miska S (2009) Effects of boundary conditions and friction on static buckling of pipe in a horizontal well. SPE J 14(4): 782–796, SPE 111511-PA. http://dx.doi.org/10.2118/111511-PA

Gao G, Miska S (2010a) Dynamic buckling and snaking motion of rotating drilling pipe in a horizontal well. SPE J 15(3): 867–877, SPE 113883-PA. http://dx.doi.org/10.2118/113883-PA

Gao G, Miska S (2010b) Effects of friction on post-buckling behavior and axial load transfer in a horizontal well. In: SPE Production and Operations Symposium, 4–8 April, Oklahoma City, Oklahoma, SPE 120084-MS. http://dx.doi.org/10.2118/120084-MS

Han SM, Benaroya H (2002) Nonlinear and stochastic dynamics of compliant offshore structures. Solid mechanics and its applications. Springer Science + Business Media, Dordrecht, 274 pp

He X, Kyllingstad A (1995) Helical buckling and lock-up conditions for coiled tubing in curved wells. SPE Drill Completion 10(1): 10–15, SPE 25370-PA. http://dx.doi.org/10.2118/25370-PA

Huang W, Gao D, Wei S, Chen P (2015a) Boundary conditions a key factor in tubular-string buckling. SPE J 20(6) 1409–1420, SPE 174087-PA. http://dx.doi.org/10.2118/174087-PA

Huang W, Gao D, Liu F (2015b) Buckling analysis of tubular strings in horizontal wells. SPE J 20(2): 405–416, SPE 171551-PA. http://dx.doi.org/10.2118/171551-PA

Lubinski A, Althouse WS, Logan JJ (1962) Helical buckling of tubing sealed in packers. Petroleum Trans 14(6): 655–670, SPE 178-PA. http://dx.doi.org/10.2118/178-PA

Miska S, Cunha JC (1995) An analysis of helical buckling of tubulars subjected to axial and torsional loading in inclined wellbores. In: SPE Production Operations Symposium, 2–4 April, Oklahoma City, Oklahoma, SPE 29460-MS. http://dx.doi.org/10.2118/29460-MS

Mitchell RF (1986) Simple frictional analysis of helical buckling of tubing. SPE Drill Eng 1(6): 457–465, SPE 13064-PA. http://dx.doi.org/10.2118/13064-PA

Mitchell RF (1988) New concepts for helical buckling. SPE Drill Eng 3(3) 303–310, SPE 15470-PA. http://dx.doi.org/10.2118/15470-PA

Mitchell RF (1996a) Buckling analysis in deviated wells—a practical method. In SPE Annual Technical Conference and Exhibition, 6–9 October, Denver, Colorado, SPE 36761-MS. http://dx.doi.org/10.2118/36761-MS

Mitchell RF (1996b) Comprehensive Analysis of Buckling with Friction. SPE Drill Completion 11(3): 178–184, SPE 29457-PA. http://dx.doi.org/10.2118/29457-PA

Mitchell RF (1997) Effects of well deviation on helical buckling. SPE Drill Completion 12(1): 63–70, SPE 29462-PA. http://dx.doi.org/10.2118/29462-PA

Mitchell RF (1999) A buckling criterion for constant-curvature wellbores. SPE J 4(4): 349–352, SPE 57896-PA. http://dx.doi.org/10.2118/57896-PA

Mitchell RF (2002) Exact analytic solutions for pipe buckling in vertical and horizontal wells. SPE J 7(4): 373–390, SPE 72079-PA. http://dx.doi.org/10.2118/72079-PA

Mitchell RF (2007) The Effect of friction on initial buckling of tubing and flowlines. SPE Drill Completions 22(2): 112–118, SPE 99099-PA. http://dx.doi.org/10.2118/99099-PA

Mitchell RF (2008) Tubing buckling—the state of the art. SPE Drill Completions 23(4), SPE 104267-PA. http://dx.doi.org/10.2118/104267-PA

Niedzwecki JM, Thampi SK (1988) Heave response of long riserless drill strings. Ocean Eng 15(5):457–469. https://doi.org/10.1016/0029-8018(88)90010-8

Park HI, Hong YP, Nakamura M, Koterayama W (2002) An experimental study on transverse vibrations of a highly flexible free-hanging pipe in water. In: Proceedings of the Twelfth International Offshore and Polar Engineering Conference, 26–31 May, Kitakyushu, Japan

Paslay PR, Bogy DB (1964) The stability of a circular rod laterally constrained to be in contact with an inclined circular cylinder. ASME J Appl Mech 31(4):605–610. https://doi.org/10.1115/1.3629721

Qiu W, Miska S, Volk L (1998) Drill pipe and coiled tubing buckling analysis in a hole of constant curvature. In: SPE Permian Basin Oil and Gas Recovery Conference, 23–26 March, Midland, Texas, SPE 39795-MS. http://dx.doi.org/10.2118/39795-MS

Qiu W, Miska SZ, Volk LJ (1999) Effect of coiled-tubing initial configuration on buckling behavior in a constant-curvature hole. SPE J 4(1): 64–71, SPE 55682-PA. http://dx.doi.org/10.2118/55682-PA

Rao SS (2007) Vibration of continuous systems. Wiley, Hoboken, 744 pp

Rocha LAS, Azuaga D, Andrade R, Vieira JLB, Santos OLA (2006) Perfuração Direcional (Directional Drilling). Rio de Janeiro: Editora Interciência, 342 pp

Saliés JB (1994) Experimental Study and Mathematical Modeling of Helical Buckling of Tubulars in Inclined Wellbores. PhD Thesis, The University of Tulsa, Tulsa

Sparks CP (2002) Transverse modal vibrations of vertical tensioned risers—a simplified analytical approach. Oil Gas Sci Technol 57(1):71–86. https://doi.org/10.2516/ogst:2002005

Sparks CP, Cabillic JP, Schawann J (1982) Longitudinal resonant behavior of very deep water risers. In: Offshore Technology Conference, 3–6 May, Houston, Texas. OTC 4317-MS. http://dx.doi.org/10.4043/4317-MS

Sun Y, Yu Y, Liu B (2014) Closed form solutions for predicting static and dynamic buckling behaviors of a drillstring in a horizontal well. Eur J Mech A/Solids 49:362–372. https://doi.org/10.1016/j.euromechsol.2014.08.008

Timoshenko S (1937) Vibration problems in engineering. D. Van Nostrand Company, Inc., Princeton, 476 pp

Wicks N, Wardle BL, Pafitis D (2007) Horizontal cylinder-in-cylinder buckling under compression and torsion: review and application to composite drill pipe. Int J Mech Sci 50(3):538–549. https://doi.org/10.1016/j.ijmecsci.2007.08.005

Chapter 3
Models for Dynamic Column Buckling

In order to fulfill the proposed objectives, we describe mathematical models on this section. The procedure used to create the model was incremental: we developed four models, with each model pushing the previous one a step further.

The solution starts with Model I, which is the same as proposed by Gao and Miska (2010). This is considered the base model, since it is the most simplified one. On this model, there is no friction, the well segment is always horizontal and the boundary at $x = 0$ is fixed. Improving this model there is Model II, which considers the friction force—but the segment is still horizontal and the boundary at $x = 0$ is still fixed. It is worth noting that Model II is the minimum requirement to verify the hypothesis that the friction force is different during tripping in and tripping out. Moving further, Model III considers the well inclination as well; therefore, any well trajectory can be studied, as long as the angle at each depth is provided. Finally, Model IV considers a periodic excitation at the boundary $x = 0$. This is a necessary improvement to consider a sea environment, since the column is subjected to a heave motion caused by the vessel heave motion; Models I to III can be applied only to onshore wells, where the column does not suffer any kind of periodic motion. Finally, while all models are subdivided into a tripping in case and a tripping out case, the column is not actually moving forward or backward; all models consider a fixed length of column vibrating around its equilibrium position for that very specific length, but under different hypotheses depending if the column is on a tripping in case or on a tripping out case. Table 3.1 sums up the hypotheses of all models.

3.1 Model I: Columns Without Friction

This model uses Figs. 2.10 and 2.11 as its basis, with the basic concepts covered in Sect. 2.3. On the following model, only a horizontal segment of well is considered. Both the well radius and the column radius are considered constant for the whole horizontal segment and the clearance between the two radii is considered small. There are some simplifications regarding the loads on the column as well: the effect of viscous damping was neglected and there is no imposed torque on the two ends, while the column rotary speed is constant. Lastly, as mentioned before for this model, the

© The Author(s) 2018

M. A. Jaculli and J. R. P. Mendes, *Dynamic Buckling of Columns Inside Oil Wells*,
SpringerBriefs in Petroleum Geoscience & Engineering,
https://doi.org/10.1007/978-3-319-91208-0_3

Table 3.1 Summary of hypothesis for all four models

	Friction force	Slant segments	Periodic motion on boundary
Model I			
Model II	X		
Model III	X	X	
Model IV	X	X	X

effect of friction is neglected. The idea behind using this model, despite it neglecting the friction force, is to do an initial observation regarding the effects of buckling on the column dynamic behavior, especially the contact force between the column and the well.

Finally, it is worth pointing that despite the text referring to the internal cylindrical element as "column" and the external cylindrical element as "well", the model is not restricted to only this scenario. As mentioned before, several operations involve the use of columns inside another column, such as lowering a tubing string inside a cased hole; a coiled tubing string inside a tubing string; or a sand screen using a work string inside an open hole. Therefore, usage of terms "column" and "well" is only to improve understanding. Lastly, the model by Gao and Miska (2010) was developed for drill strings, which rotate while moving forward. This does not happen in completion scenarios; however, the effect of rotation is kept, thus the model can still be of use for analyzing drill strings.

3.1.1   Model for Tripping in

During tripping in, the column will be subjected to compressive loads which will cause buckling. Therefore, the point C_0 from Fig. 2.11a which is initially on the lower portion of the well with coordinates (x, 0, -r) will displace to the position of point C from Fig. 2.11b with coordinates $(x+u_x, r*\sin\theta, -r*\cos\theta)$ on a certain time t. To keep the sign convention consistent, the displacement u_x is added up, despite being negative since it is a contraction. This displacement includes the effects of axial contraction u_a and bending contraction u_b, as explained beforehand. The coordinate x is the initial position along the horizontal segment of well, the coordinate θ is the angle defined between the z axis and the normal vector $\hat{p}$ and the distance r is the difference between the well radius and the column radius—also known as clearance.

Starting from the origin defined at (0, 0, 0), the position vector $\vec{r}$ between point C and this origin is given by Eq. 3.1:

$$\vec{r}(x, t) = (x + u_x)\hat{i} + r\sin\theta\,\hat{j} - r\cos\theta\,\hat{k} \tag{3.1}$$

The unitary vector $\vec{\tau}$, tangential to the column axial axis, is defined by Eq. 3.2:

$$\vec{\tau}(x,t) = \frac{1}{c}\frac{\partial \vec{r}}{\partial x} \tag{3.2}$$

where c is the vector norm of $\partial \vec{r}/\partial x$ Developing Eq. 3.2:

$$\vec{\tau}(x,t) = \frac{1}{c}\frac{\partial}{\partial x}\left[(x+u_x)\hat{i} + r\sin\theta\,\hat{j} - r\cos\theta\,\hat{k}\right] \tag{3.3}$$

$$\vec{\tau}(x,t) = \frac{1}{c}\left[\left(1+\frac{\partial u_x}{\partial x}\right)\hat{i} + r\cos\theta\frac{\partial\theta}{\partial x}\hat{j} + r\sin\theta\frac{\partial\theta}{\partial x}\hat{k}\right] \tag{3.4}$$

It is noted that the norm c is given by $\sqrt{1+r^2\frac{\partial^2\theta}{\partial x^2}}$. Considering that r is very small (r ≪ 1), then c ≈ 1. Also, $\frac{\partial u_x}{\partial x} \ll 1$, therefore this term can be neglected on the component $\hat{i}$. Introducing the unitary vector $\hat{q}$, corresponding to the tangential direction on the contact as seen on Fig. 2.10, defined by:

$$\hat{q} = \cos\theta\,\hat{j} + \sin\theta\,\hat{k} \tag{3.5}$$

Equation 3.4 can be simplified by using Eq. 3.5:

$$\vec{\tau}(x,t) = \hat{i} + r\frac{\partial\theta}{\partial x}\hat{q} \tag{3.6}$$

As said before, Eq. 3.5 provides the unitary vector $\vec{\tau}$, which is tangential to the column axial axis for each coordinate x. In order to find the normal unitary vector to $\vec{\tau}$, namely $\vec{n}$, the derivative of $\vec{\tau}$ is taken, according to Eq. 3.7:

$$\vec{n} = \frac{1}{k}\frac{\partial\vec{\tau}}{\partial x} \tag{3.7}$$

where k is vector norm of $\partial\vec{\tau}/\partial x$. Taking the derivative of Eq. 3.4:

$$\frac{\partial\vec{\tau}}{\partial x} = \frac{\partial}{\partial x}\left[\hat{i} + r\cos\theta\frac{\partial\theta}{\partial x}\hat{j} + r\sin\theta\frac{\partial\theta}{\partial x}\hat{k}\right] \tag{3.8}$$

$$\frac{\partial\vec{\tau}}{\partial x} = -r\sin\theta\left(\frac{\partial\theta}{\partial x}\right)^2\hat{j} + r\cos\theta\frac{\partial^2\theta}{\partial x^2}\hat{j} + r\cos\theta\left(\frac{\partial\theta}{\partial x}\right)^2\hat{k} + r\sin\theta\frac{\partial^2\theta}{\partial x^2}\hat{k} \tag{3.9}$$

Introducing the unitary vector $\hat{p}$, corresponding to the normal direction on the contact as seen on Fig. 2.10 and defined by:

$$\hat{p} = \sin\theta\,\hat{j} - \cos\theta\,\hat{k} \tag{3.10}$$

Substituting Eqs. 3.5 and 3.10 into Eq. 3.9:

$$\frac{\partial\vec{\tau}}{\partial x} = -r\left(\frac{\partial\theta}{\partial x}\right)^2\hat{p} + r\frac{\partial^2\theta}{\partial x^2}\hat{q} \tag{3.11}$$

$$\frac{\partial \vec{\tau}}{\partial x} = k_r \hat{p} + k_\theta \hat{q} \tag{3.12}$$

where k_r and k_θ on Eq. 3.12 are given by:

$$k_r = -r\left(\frac{\partial \theta}{\partial x}\right)^2 \tag{3.13}$$

$$k_\theta = r\frac{\partial^2 \theta}{\partial x^2} \tag{3.14}$$

Thus, the modulus k from Eq. 3.7 is given by:

$$k = \sqrt{k_r^2 + k_\theta^2} \tag{3.15}$$

Therefore, Eq. 3.7 becomes:

$$k\vec{n} = k_r \hat{p} + k_\theta \hat{q} \tag{3.16}$$

Lastly, the binormal unitary vector $\vec{b}$—which is perpendicular to both $\vec{\tau}$ and $\vec{n}$—is defined by Eq. 3.17:

$$\vec{b} = \vec{\tau} \times \vec{n} \tag{3.17}$$

or alternatively:

$$k\vec{b} = \vec{\tau} \times k\vec{n} \tag{3.18}$$

Substituting Eqs. 3.4 and 3.9 into Eq. 3.18:

$$
\begin{aligned}
k\vec{b} = {} & \left(\hat{i} + r\cos\theta\frac{\partial\theta}{\partial x}\hat{j} + r\sin\theta\frac{\partial\theta}{\partial x}\hat{k}\right) \\
& \times \left(-r\sin\theta\left(\frac{\partial\theta}{\partial x}\right)^2\hat{j} + r\cos\theta\frac{\partial^2\theta}{\partial x^2}\hat{j} + r\cos\theta\left(\frac{\partial\theta}{\partial x}\right)^2\hat{k} + r\sin\theta\frac{\partial^2\theta}{\partial x^2}\hat{k}\right)
\end{aligned}
\tag{3.19}
$$

$$
\begin{aligned}
k\vec{b} = {} & r\cos\theta\frac{\partial\theta}{\partial x}r\cos\theta\left(\frac{\partial\theta}{\partial x}\right)^2\hat{i} + r\cos\theta\frac{\partial\theta}{\partial x}r\sin\theta\frac{\partial^2\theta}{\partial x^2}\hat{i} \\
& + r\sin\theta\frac{\partial\theta}{\partial x}r\sin\theta\left(\frac{\partial\theta}{\partial x}\right)^2\hat{i} - r\sin\theta\frac{\partial\theta}{\partial x}r\cos\theta\frac{\partial^2\theta}{\partial x^2}\hat{i} \\
& - r\cos\theta\left(\frac{\partial\theta}{\partial x}\right)^2\hat{j} - r\sin\theta\frac{\partial^2\theta}{\partial x^2}\hat{j} - r\sin\theta\left(\frac{\partial\theta}{\partial x}\right)^2\hat{k} + r\cos\theta\frac{\partial^2\theta}{\partial x^2}\hat{k}
\end{aligned}
\tag{3.20}
$$

$$
\begin{aligned}
k\vec{b} = {} & r^2\cos^2\theta\left(\frac{\partial\theta}{\partial x}\right)^3\hat{i} + r^2\sin^2\theta\left(\frac{\partial\theta}{\partial x}\right)^3\hat{i} - r\cos\theta\left(\frac{\partial\theta}{\partial x}\right)^2\hat{j} \\
& - r\sin\theta\frac{\partial^2\theta}{\partial x^2}\hat{j} - r\sin\theta\left(\frac{\partial\theta}{\partial x}\right)^2\hat{k} + r\cos\theta\frac{\partial^2\theta}{\partial x^2}\hat{k}
\end{aligned}
\tag{3.21}
$$

$$k\vec{b} = \left[r^2 \left(\frac{\partial \theta}{\partial x} \right)^3 + r^2 \left(\frac{\partial \theta}{\partial x} \right)^3 \right] \hat{i} - r \frac{\partial^2 \theta}{\partial x^2} \hat{p} - r \left(\frac{\partial \theta}{\partial x} \right)^2 \hat{q} \tag{3.22}$$

The term of component $\hat{i}$ can be neglected since it is too small when compared with the terms from components $\hat{p}$ and $\hat{q}$, since the value of r is very small. Using Eqs. 3.13–3.14 into Eq. 3.22:

$$k\vec{b} = -k_\theta \hat{p} + k_r \hat{q} \tag{3.23}$$

Up to this point, the derivatives of the position vector $\vec{r}$ were taken in respect to space. Since this is a dynamic problem, derivatives in respect to time must be taken as well. The velocity vector $\vec{v}$ will be given by Eq. 3.24:

$$\vec{v}(x, t) = \frac{\partial \vec{r}}{\partial t} \tag{3.24}$$

Substituting Eq. 3.1 into Eq. 3.24:

$$\vec{v}(x, t) = \frac{\partial}{\partial t} \left[(x + u_x) \hat{i} + r \sin \theta \, \hat{j} - r \cos \theta \, \hat{k} \right] \tag{3.25}$$

$$\vec{v}(x, t) = \frac{\partial u_x}{\partial t} \hat{i} + r \cos \theta \frac{\partial \theta}{\partial t} \hat{j} + r \sin \theta \frac{\partial \theta}{\partial t} \hat{k} \tag{3.26}$$

$$\vec{v}(x, t) = \frac{\partial u_x}{\partial t} \hat{i} + r \frac{\partial \theta}{\partial t} \hat{q} \tag{3.27}$$

Both Eq. 3.5 and the knowledge that $\partial x / \partial t = 0$, since x is an independent coordinate, were used. The acceleration vector $\vec{a}$ will be given by Eq. 3.28:

$$\vec{a}(x, t) = \frac{\partial \vec{v}}{\partial t} \tag{3.28}$$

Substituting Eq. 3.26 into Eq. 3.28:

$$\vec{a}(x, t) = \frac{\partial}{\partial t} \left[\frac{\partial u_x}{\partial t} \hat{i} + r \cos \theta \frac{\partial \theta}{\partial t} \hat{j} + r \sin \theta \frac{\partial \theta}{\partial t} \hat{k} \right] \tag{3.29}$$

$$\vec{a}(x, t) = \frac{\partial^2 u_x}{\partial t^2} \hat{i} + \left[-r \sin \theta \left(\frac{\partial \theta}{\partial t} \right)^2 + r \cos \theta \frac{\partial^2 \theta}{\partial t^2} \right] \hat{j} + \left[r \cos \theta \left(\frac{\partial \theta}{\partial t} \right)^2 + r \sin \theta \frac{\partial^2 \theta}{\partial t^2} \right] \hat{k} \tag{3.30}$$

$$\vec{a}(x, t) = \frac{\partial^2 u_x}{\partial t^2} \hat{i} - r \left(\frac{\partial \theta}{\partial t} \right)^2 \hat{p} + r \frac{\partial^2 \theta}{\partial t^2} \hat{q} \tag{3.31}$$

Equations 3.5 and 3.10 were used. Since the problem is also related to the angular rotation of the column, it is necessary to define an angular velocity vector $\vec{\Omega}$, which has contributions from both the column own rotary speed ω as well as the change in direction given by the derivative of $\vec{\tau}$:

$$\vec{\Omega}(x, t) = \omega\vec{\tau} + \frac{\partial\vec{\tau}}{\partial t} \tag{3.32}$$

Substituting Eq. 3.4 into Eq. 3.32:

$$\vec{\Omega}(x, t) = \omega\left[\hat{i} + r\cos\theta\frac{\partial\theta}{\partial x}\hat{j} + r\sin\theta\frac{\partial\theta}{\partial x}\hat{k}\right] + \frac{\partial}{\partial t}\left[\hat{i} + r\cos\theta\frac{\partial\theta}{\partial x}\hat{j} + r\sin\theta\frac{\partial\theta}{\partial x}\hat{k}\right] \tag{3.33}$$

$$\vec{\Omega}(x, t) = \omega\hat{i} + \omega r\cos\theta\frac{\partial\theta}{\partial x}\hat{j} + \omega r\sin\theta\frac{\partial\theta}{\partial x}\hat{k} - r\sin\theta\frac{\partial\theta}{\partial t}\frac{\partial\theta}{\partial x}\hat{j}$$
$$+ r\cos\theta\frac{\partial^2\theta}{\partial x\partial t}\hat{j} + r\cos\theta\frac{\partial\theta}{\partial t}\frac{\partial\theta}{\partial x}\hat{k} + r\sin\theta\frac{\partial^2\theta}{\partial x\partial t}\hat{k} \tag{3.34}$$

$$\vec{\Omega}(x, t) = \omega\hat{i} + \omega r\frac{\partial\theta}{\partial x}\hat{q} - r\frac{\partial\theta}{\partial t}\frac{\partial\theta}{\partial x}\hat{p} + r\frac{\partial^2\theta}{\partial x\partial t}\hat{q} \tag{3.35}$$

$$\vec{\Omega}(x, t) = \omega\hat{i} + \omega_r\hat{p} + \left(\omega r\frac{\partial\theta}{\partial x} + \omega_\theta\right)\hat{q} \tag{3.36}$$

where ω_r and ω_θ on Eq. 3.36 are given by:

$$\omega_r = -r\frac{\partial\theta}{\partial t}\frac{\partial\theta}{\partial x} \tag{3.37}$$

$$\omega_\theta = r\frac{\partial^2\theta}{\partial x\partial t} \tag{3.38}$$

On Eq. 3.36, the terms of coordinates $\hat{p}$ and $\hat{q}$ will generate angular moments much smaller than the angular moment from coordinate $\hat{i}$ and thus can be neglected. Finally, in order to relate position, linear velocity, angular velocity and acceleration with forces and moments, it is necessary to calculate the linear and angular momentums of an infinitesimal column element with length dx. Firstly, the linear momentum:

$$\vec{P}(x, t) = m_p\vec{v}dx \tag{3.39}$$

where m_p is the column mass per unit of length. Substituting Eq. 3.27 into Eq. 3.39:

$$\vec{P}(x, t) = m_p\frac{\partial u_x}{\partial t}dx\,\hat{i} + m_p r\frac{\partial\theta}{\partial t}dx\,\hat{q} \tag{3.40}$$

Then, the angular moment:

$$\vec{H}_0(x, t) = \left(m_p\vec{r}\times\vec{v} + I_p\omega\vec{\tau}\right)dx \tag{3.41}$$

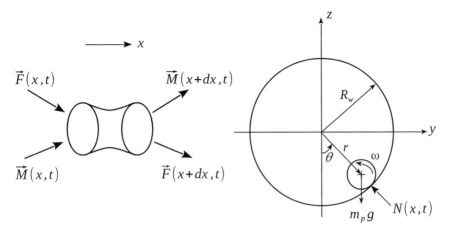

Fig. 3.1 Loads acting on the column. On the left side, the internal forces and moments acting on an infinitesimal element. On the right one, the weight and the normal contact force

where I_p is the mass moment of inertia per unit of length and is related with the area moment of inertia I through the expression $I_p = 2\rho I$, where ρ is the material density (in kg/m³). The cross product $\vec{r} \times \vec{v}$ is too small when compared to the angular moment generated by ω and can be neglected. Thus, neglecting this term and substituting Eq. 3.4 into Eq. 3.41:

$$\vec{H}_0(x, t) = \left[I_p\omega\left(\hat{i} + r\cos\theta\frac{\partial\theta}{\partial x}\hat{j} + r\sin\theta\frac{\partial\theta}{\partial x}\hat{k} \right) \right]dx \tag{3.42}$$

$$\vec{H}_0(x, t) = \left[I_p\omega\hat{i} + I_p\omega r\frac{\partial\theta}{\partial x}\hat{q} \right]dx \tag{3.43}$$

The loads acting on the column can be seen on Fig. 3.1. The column will be subjected to internal forces and internal moments due to axial tension/compression and bending, respectively, and also its own weight and the normal contact force with respect to the wellbore.

The internal force $\vec{F}$ can be written as a function of either the coordinates $\hat{i}, \hat{j}$ and $\hat{k}$ or $\hat{i}, \hat{p}$ and $\hat{q}$. For convenience, since the vector decomposition will be made later on the $\hat{i}, \hat{p}$ and $\hat{q}$ coordinates, $\vec{F}$ is defined as being:

$$\vec{F} = F_x\hat{i} + F_r\hat{p} + F_\theta\hat{q} \tag{3.44}$$

where F_x, F_r and F_θ are the components of $\vec{F}$. To calculate the space derivative of $\vec{F}$, it can be more interesting to return to coordinates $\hat{i}, \hat{j}$ and $\hat{k}$. Substituting Eqs. 3.5 and 3.10 into Eq. 3.44:

$$\vec{F} = F_x\hat{i} + F_r\left(\sin\theta\hat{j} - \cos\theta\hat{k} \right) + F_\theta\left(\cos\theta\hat{j} + \sin\theta\hat{k} \right) \tag{3.45}$$

$$\overrightarrow{F} = F_x\hat{i} + (F_r \sin\theta + F_\theta \cos\theta)\hat{j} + (-F_r \cos\theta + F_\theta \sin\theta)\hat{k} \qquad (3.46)$$

Taking the spatial derivative of Eq. 3.46:

$$\frac{\partial \overrightarrow{F}}{\partial x} = \frac{\partial}{\partial x}\left[F_x\hat{i} + (F_r \sin\theta + F_\theta \cos\theta)\hat{j} + (-F_r \cos\theta + F_\theta \sin\theta)\hat{k} \right] \qquad (3.47)$$

$$\frac{\partial \overrightarrow{F}}{\partial x} = \frac{\partial F_x}{\partial x}\hat{i} + \frac{\partial F_r}{\partial x}\sin\theta\,\hat{j} + F_r \cos\theta\frac{\partial\theta}{\partial x}\hat{j} + \frac{\partial F_\theta}{\partial x}\cos\theta\,\hat{j} - F_\theta \sin\theta\frac{\partial\theta}{\partial x}\hat{j}$$
$$- \frac{\partial F_r}{\partial x}\cos\theta\,\hat{k} + F_r \sin\theta\frac{\partial\theta}{\partial x}\hat{k} + \frac{\partial F_\theta}{\partial x}\sin\theta\,\hat{k} + F_\theta \cos\theta\frac{\partial\theta}{\partial x}\hat{k} \qquad (3.48)$$

Substituting Eqs. 3.5 and 3.10 once again:

$$\frac{\partial \overrightarrow{F}}{\partial x} = \frac{\partial F_x}{\partial x}\hat{i} + \left(\frac{\partial F_r}{\partial x} - F_\theta\frac{\partial\theta}{\partial x}\right)\hat{p} + \left(\frac{\partial F_\theta}{\partial x} + F_r\frac{\partial\theta}{\partial x}\right)\hat{q} \qquad (3.49)$$

Similarly to the internal force $\overrightarrow{F}$, the internal moment $\overrightarrow{M}$ can be written on both sets of coordinates. There is no contribution on $\hat{i}$ since there is no applied torque on the column. Therefore:

$$\overrightarrow{M} = M_r\hat{p} + M_\theta\hat{q} \qquad (3.50)$$

where M_r and M_θ are the components of $\overrightarrow{M}$. Again, using the $\hat{i}$, $\hat{j}$ and $\hat{k}$ coordinates to calculate the spatial derivative:

$$\overrightarrow{M} = M_r\left(\sin\theta\,\hat{j} - \cos\theta\,\hat{k}\right) + M_\theta\left(\cos\theta\,\hat{j} + \sin\theta\,\hat{k}\right) \qquad (3.51)$$

$$\overrightarrow{M} = (M_r \sin\theta + M_\theta \cos\theta)\hat{j} + (-M_r \cos\theta + M_\theta \sin\theta)\hat{k} \qquad (3.52)$$

Taking the spatial derivative of Eq. 3.52:

$$\frac{\partial \overrightarrow{M}}{\partial x} = \frac{\partial}{\partial x}\left[(M_r \sin\theta + M_\theta \cos\theta)\hat{j} + (-M_r \cos\theta + M_\theta \sin\theta)\hat{k} \right] \qquad (3.53)$$

$$\frac{\partial \overrightarrow{M}}{\partial x} = \frac{\partial M_r}{\partial x}\sin\theta\,\hat{j} + M_r \cos\theta\frac{\partial\theta}{\partial x}\hat{j} + \frac{\partial M_\theta}{\partial x}\cos\theta\,\hat{j} - M_\theta \sin\theta\frac{\partial\theta}{\partial x}\hat{j}$$
$$- \frac{\partial M_r}{\partial x}\cos\theta\,\hat{k} + M_r \sin\theta\frac{\partial\theta}{\partial x}\hat{k} + \frac{\partial M_\theta}{\partial x}\sin\theta\,\hat{k} + M_\theta \cos\theta\frac{\partial\theta}{\partial x}\hat{k} \qquad (3.54)$$

Substituting again Eqs. 3.5 and 3.10:

$$\frac{\partial \overrightarrow{M}}{\partial x} = \left(\frac{\partial M_r}{\partial x} - M_\theta\frac{\partial\theta}{\partial x}\right)\hat{p} + \left(\frac{\partial M_\theta}{\partial x} + M_r\frac{\partial\theta}{\partial x}\right)\hat{q} \qquad (3.55)$$

The column own weight per unit of length $\vec{q}_p$ acts on the central axis and on the negative direction of the z axis. Therefore:

$$\vec{q}_p = -m_p g dx \hat{k} \tag{3.56}$$

where, once more, m_p is the column mass per unit of length, g is the gravitational acceleration and dx is the length of an infinitesimal element. Changing to $\hat{p}$ and $\hat{q}$ coordinates once more:

$$\vec{q}_p = m_p g dx \cos \theta \, \hat{p} - m_p g dx \sin \theta \, \hat{q} \tag{3.57}$$

It can be noted that $\hat{k} = \cos \theta \, \hat{p} - \sin \theta \, \hat{q}$ by using Eqs. 3.5 and 3.10. The normal contact force is already aligned to $\hat{p}$ but on the opposite direction defined on Fig. 2. 10. Therefore:

$$\overrightarrow{N} = -N dx \hat{p} \tag{3.58}$$

where N is the normal contact force per unit of length—thus having units of N/m. The total external force per unit of length is obtained by adding them up:

$$\vec{f} = \frac{\vec{q}_p}{dx} + \frac{\overrightarrow{N}}{dx} \tag{3.59}$$

Substituting Eqs. 3.57–3.58 into Eq. 3.59:

$$\vec{f} = (m_p g \cos \theta - N)\hat{p} - m_p g \sin \theta \, \hat{q} \tag{3.60}$$

From the Strength of Materials, it is known that internal forces and internal moments are directly tied to displacements and strains. Defining the total axial displacement u_x as being:

$$u_x = u_a + u_b \tag{3.61}$$

where u_a is the axial displacement caused by axial tension and compression and u_b is the axial displacement caused due to bending. The displacement u_a comes from Hooke's Law:

$$F_x(x, t) = -EA \frac{\partial u_a}{\partial x} \tag{3.62}$$

where E is the material Young's modulus and A is the cross sectional area. The displacement u_b can be obtained by:

$$u_b = -\frac{1}{2} r^2 \int_0^x \left(\frac{\partial \theta}{\partial x} \right)^2 dx \tag{3.63}$$

Taking the spatial derivative of Eq. 3.61 and substituting Eqs. 3.62–3.63, with r constant:

$$\frac{\partial u_x}{\partial x} = \frac{\partial u_a}{\partial x} + \frac{\partial u_b}{\partial x} \tag{3.64}$$

$$\frac{\partial u_x}{\partial x} = -\frac{F_x(x, t)}{EA} - \frac{1}{2}r^2\left(\frac{\partial \theta}{\partial x}\right)^2 \tag{3.65}$$

$$F_x(x, t) = -EA\frac{\partial u_x}{\partial x} - \frac{1}{2}EAr^2\left(\frac{\partial \theta}{\partial x}\right)^2 \tag{3.66}$$

Meanwhile, the bending moment can be obtained through its relation with the curvature radius, in this case given by the binormal vector:

$$\overrightarrow{M}(x, t) = -EIk\vec{b} \tag{3.67}$$

where I is the area moment of inertia. Substituting Eq. 3.22 into Eq. 3.67:

$$\overrightarrow{M}(x, t) = -EI\left[-r\frac{\partial^2\theta}{\partial x^2}\hat{p} - r\left(\frac{\partial \theta}{\partial x}\right)^2\hat{q}\right] \tag{3.68}$$

$$\overrightarrow{M}(x, t) = EIr\frac{\partial^2\theta}{\partial x^2}\hat{p} + EIr\left(\frac{\partial \theta}{\partial x}\right)^2\hat{q} \tag{3.69}$$

Comparing Eq. 3.69 with Eq. 3.50, it can be concluded that:

$$M_r = EIr\frac{\partial^2\theta}{\partial x^2} \tag{3.70}$$

$$M_\theta = EIr\left(\frac{\partial \theta}{\partial x}\right)^2 \tag{3.71}$$

Now that the loads and linear and angular momentums were defined, it is time to apply Newton's Second Law to find the motion equations for the column. Starting with the linear momentum:

$$\sum \overrightarrow{F}_i = \frac{\partial \overrightarrow{P}}{\partial t} \tag{3.72}$$

Substituting Eq. 3.39 into Eq. 3.72 and introducing the loads defined previously according to the convention of Fig. 3.1:

$$\overrightarrow{F} - \left(\overrightarrow{F} + \frac{\partial \overrightarrow{F}}{\partial x}dx\right) + \vec{f}dx = m_p\frac{\partial \vec{v}}{\partial t}dx \tag{3.73}$$

Simplifying Eq. 3.73:

$$\frac{\partial \overrightarrow{F}}{\partial x} - \overrightarrow{f} + m_p \frac{\partial \overrightarrow{v}}{\partial t} = 0 \tag{3.74}$$

Substituting Eqs. 3.31, 3.49 and 3.60 into Eq. 3.74:

$$\frac{\partial F_x}{\partial x}\hat{i} + \left(\frac{\partial F_r}{\partial x} - F_\theta \frac{\partial \theta}{\partial x}\right)\hat{p} + \left(\frac{\partial F_\theta}{\partial x} + F_r \frac{\partial \theta}{\partial x}\right)\hat{q} - \left(m_p g \cos \theta - N\right)\hat{p}$$
$$+ m_p g \sin \theta \, \hat{q} + m_p \left[\frac{\partial^2 u_x}{\partial t^2}\hat{i} - r\left(\frac{\partial \theta}{\partial t}\right)^2 \hat{p} + r\frac{\partial^2 \theta}{\partial t^2}\hat{q}\right] = 0 \tag{3.75}$$

$$\left(\frac{\partial F_x}{\partial x} + m_p \frac{\partial^2 u_x}{\partial t^2}\right)\hat{i}$$
$$+ \left(\frac{\partial F_r}{\partial x} - F_\theta \frac{\partial \theta}{\partial x} + N - m_p g \cos \theta - m_p r \left(\frac{\partial \theta}{\partial t}\right)^2\right)\hat{p}$$
$$+ \left(\frac{\partial F_\theta}{\partial x} + F_r \frac{\partial \theta}{\partial x} + m_p g \sin \theta + m_p r \frac{\partial^2 \theta}{\partial t^2}\right)\hat{q} = 0 \tag{3.76}$$

Separating Eq. 3.76 into components $\hat{i}$, $\hat{p}$ and $\hat{q}$:

$$\frac{\partial F_x}{\partial x} + m_p \frac{\partial^2 u_x}{\partial t^2} = 0 \tag{3.77}$$

$$\frac{\partial F_r}{\partial x} - F_\theta \frac{\partial \theta}{\partial x} + N - m_p g \cos \theta - m_p r \left(\frac{\partial \theta}{\partial t}\right)^2 = 0 \tag{3.78}$$

$$\frac{\partial F_\theta}{\partial x} + F_r \frac{\partial \theta}{\partial x} + m_p g \sin \theta + m_p r \frac{\partial^2 \theta}{\partial t^2} = 0 \tag{3.79}$$

Substituting Eq. 3.66 into Eq. 3.77:

$$\frac{\partial}{\partial x}\left[-EA\frac{\partial u_x}{\partial x} - \frac{1}{2}EAr^2\left(\frac{\partial \theta}{\partial x}\right)^2\right] + m_p \frac{\partial^2 u_x}{\partial t^2} = 0 \tag{3.80}$$

Manipulating Eq. 3.80, a motion equation relating the axial displacement u_x and the angular displacement θ is found:

$$EA\frac{\partial^2 u_x}{\partial x^2} - m_p \frac{\partial^2 u_x}{\partial t^2} + EAr^2 \frac{\partial \theta}{\partial x}\frac{\partial^2 \theta}{\partial x^2} = 0 \tag{3.81}$$

Equation 3.81 still has two unknowns; another equation is needed to calculate the two displacements. It can be obtained also from Newton's Second Law, but now applied to moments:

$$\sum \overrightarrow{M}_i = \frac{\partial \overrightarrow{H}_0}{\partial t} \tag{3.82}$$

Substituting the simplified Eq. 3.41 on Eq. 3.82 and introducing the loads defined previously according to the convention of Fig. 3.1:

$$\vec{M} - \left(\vec{M} + \frac{\partial \vec{M}}{\partial x}dx\right) - \vec{r} \times \vec{F} = \frac{\partial}{\partial t}\left(I_p \omega \vec{\tau} dx\right) \tag{3.83}$$

Simplifying Eq. 3.83 and taking the spatial derivative of $\vec{r}$:

$$\frac{\partial \vec{M}}{\partial x} + \vec{\tau} \times \vec{F} + I_p \omega \frac{\partial \vec{\tau}}{\partial t} = 0 \tag{3.84}$$

Substituting Eqs. 3.6, 3.44 and 3.55 and a part of Eq. 3.36 into Eq. 3.84:

$$\left(\frac{\partial M_r}{\partial x} - M_\theta \frac{\partial \theta}{\partial x}\right)\hat{p} + \left(\frac{\partial M_\theta}{\partial x} + M_r \frac{\partial \theta}{\partial x}\right)\hat{q} + \left(\hat{i} + r\frac{\partial \theta}{\partial x}\hat{q}\right)$$
$$\times \left(F_x\hat{i} + F_r\hat{p} + F_\theta\hat{q}\right) + I_p\omega\left(\omega_r\hat{p} + \omega_\theta\hat{q}\right) = 0 \tag{3.85}$$

$$\left(-F_r r\frac{\partial \theta}{\partial x}\right)\hat{i} + \left(\frac{\partial M_r}{\partial x} - M_\theta \frac{\partial \theta}{\partial x} + F_x r\frac{\partial \theta}{\partial x} - F_\theta + I_p\omega\omega_r\right)\hat{p}$$
$$+ \left(\frac{\partial M_\theta}{\partial x} + M_r \frac{\partial \theta}{\partial x} + F_r + I_p\omega\omega_\theta\right)\hat{q} = 0 \tag{3.86}$$

Separating Eq. 3.86 into components $\hat{i}$, $\hat{p}$ and $\hat{q}$:

$$-F_r r\frac{\partial \theta}{\partial x} = 0 \tag{3.87}$$

$$\frac{\partial M_r}{\partial x} - M_\theta \frac{\partial \theta}{\partial x} + F_x r\frac{\partial \theta}{\partial x} - F_\theta + I_p\omega\omega_r = 0 \tag{3.88}$$

$$\frac{\partial M_\theta}{\partial x} + M_r \frac{\partial \theta}{\partial x} + F_r + I_p\omega\omega_\theta = 0 \tag{3.89}$$

Equation 3.87 does not give much information; meanwhile, Eqs. 3.88–3.89 can be further developed. Substituting Eqs. 3.70–3.71 into Eq. 3.88:

$$\frac{\partial}{\partial x}\left[EIr\frac{\partial^2\theta}{\partial x^2}\right] - \left[EIr\left(\frac{\partial \theta}{\partial x}\right)^2\right]\frac{\partial \theta}{\partial x} + F_x r\frac{\partial \theta}{\partial x} - F_\theta + I_p\omega\omega_r = 0 \tag{3.90}$$

Isolating the F_θ component:

$$EIr\left[\frac{\partial^3\theta}{\partial x^3} - \left(\frac{\partial \theta}{\partial x}\right)^3\right] + F_x r\frac{\partial \theta}{\partial x} - F_\theta + I_p\omega\omega_r = 0 \tag{3.91}$$

$$F_\theta = EIr\left[\frac{\partial^3\theta}{\partial x^3} - \left(\frac{\partial \theta}{\partial x}\right)^3\right] + F_x r\frac{\partial \theta}{\partial x} + I_p\omega\omega_r \tag{3.92}$$

Once more, substituting Eqs. 3.70–3.71 but now on Eq. 3.89:

$$\frac{\partial}{\partial x}\left[EIr\left(\frac{\partial\theta}{\partial x}\right)^2\right] + \left[EIr\frac{\partial^2\theta}{\partial x^2}\right]\frac{\partial\theta}{\partial x} + F_r + I_p\omega\omega_\theta = 0 \tag{3.93}$$

Isolating the F_r component:

$$3EIr\frac{\partial\theta}{\partial x}\frac{\partial^2\theta}{\partial x^2} + F_r + I_p\omega\omega_\theta = 0 \tag{3.94}$$

$$F_r = -3EIr\frac{\partial\theta}{\partial x}\frac{\partial^2\theta}{\partial x^2} - I_p\omega\omega_\theta \tag{3.95}$$

Now, by knowing components F_r and F_θ, substituting Eqs. 3.92 and 3.95 into Eq. 3.78:

$$\frac{\partial}{\partial x}\left[-3EIr\frac{\partial\theta}{\partial x}\frac{\partial^2\theta}{\partial x^2} - I_p\omega\omega_\theta\right]$$
$$- \left[EIr\left[\frac{\partial^3\theta}{\partial x^3} - \left(\frac{\partial\theta}{\partial x}\right)^3\right] + F_x r\frac{\partial\theta}{\partial x} + I_p\omega\omega_r\right]\frac{\partial\theta}{\partial x}$$
$$+ N - m_p g\cos\theta - m_p r\left(\frac{\partial\theta}{\partial t}\right)^2 = 0 \tag{3.96}$$

Isolating for the normal contact force per unit of length N:

$$EIr\left[\left(\frac{\partial\theta}{\partial x}\right)^4 - 3\left(\frac{\partial^2\theta}{\partial x^2}\right)^2 - 4\frac{\partial^3\theta}{\partial x^3}\frac{\partial\theta}{\partial x}\right] - F_x r\left(\frac{\partial\theta}{\partial x}\right)^2$$
$$- I_p\omega\left[\frac{\partial\omega_\theta}{\partial x} + \omega_r\frac{\partial\theta}{\partial x}\right] + N - m_p g\cos\theta - m_p r\left(\frac{\partial\theta}{\partial t}\right)^2 = 0 \tag{3.97}$$

$$N(x,t) = -EIr\left[\left(\frac{\partial\theta}{\partial x}\right)^4 - 3\left(\frac{\partial^2\theta}{\partial x^2}\right)^2 - 4\frac{\partial^3\theta}{\partial x^3}\frac{\partial\theta}{\partial x}\right] + F_x r\left(\frac{\partial\theta}{\partial x}\right)^2$$
$$+ I_p\omega\left[\frac{\partial\omega_\theta}{\partial x} + \omega_r\frac{\partial\theta}{\partial x}\right] + m_p g\cos\theta + m_p r\left(\frac{\partial\theta}{\partial t}\right)^2 \tag{3.98}$$

Lastly, repeating the procedure for Eq. 3.79:

$$\frac{\partial}{\partial x}\left[EIr\left[\frac{\partial^3\theta}{\partial x^3} - \left(\frac{\partial\theta}{\partial x}\right)^3\right] + F_x r\frac{\partial\theta}{\partial x} + I_p\omega\omega_r\right] + \left[-3EIr\frac{\partial\theta}{\partial x}\frac{\partial^2\theta}{\partial x^2} - I_p\omega\omega_\theta\right]\frac{\partial\theta}{\partial x}$$
$$+ m_p g\sin\theta + m_p r\frac{\partial^2\theta}{\partial t^2} = 0 \tag{3.99}$$

Simplifying Eq. 3.99, an equation for the angular displacement is obtained:

$$
\mathrm{EIr}\left[\frac{\partial^4\theta}{\partial x^4} - 6\left(\frac{\partial\theta}{\partial x}\right)^2\frac{\partial^2\theta}{\partial x^2}\right] + r\left[F_x\frac{\partial^2\theta}{\partial x^2} + \frac{\partial F_x}{\partial x}\frac{\partial\theta}{\partial x}\right]
$$

$$
+ I_p\omega\left[\frac{\partial\omega_r}{\partial x} - \omega_\theta\frac{\partial\theta}{\partial x}\right] + m_pg\sin\theta + m_pr\frac{\partial^2\theta}{\partial t^2} = 0 \tag{3.100}
$$

It is possible to simplify even further Eqs. 3.98 and 3.100 by using Eqs. 3.37, 3.38 and 3.66:

$$
N(x, t) = -\mathrm{EIr}\left[\left(\frac{\partial\theta}{\partial x}\right)^4 - 3\left(\frac{\partial^2\theta}{\partial x^2}\right)^2 - 4\frac{\partial^3\theta}{\partial x^3}\frac{\partial\theta}{\partial x}\right]
$$

$$
- \mathrm{EAr}\left[\frac{\partial u_x}{\partial x}\left(\frac{\partial\theta}{\partial x}\right)^2 + \frac{1}{2}r^2\left(\frac{\partial\theta}{\partial x}\right)^4\right]
$$

$$
+ I_pr\omega\left[\frac{\partial^3\theta}{\partial x^2\partial t} - \frac{\partial\theta}{\partial t}\left(\frac{\partial\theta}{\partial x}\right)^2\right] + m_pg\cos\theta + m_pr\left(\frac{\partial\theta}{\partial t}\right)^2 \tag{3.101}
$$

$$
\mathrm{EIr}\left[\frac{\partial^4\theta}{\partial x^4} - 6\left(\frac{\partial\theta}{\partial x}\right)^2\frac{\partial^2\theta}{\partial x^2}\right] - \mathrm{EAr}\left[\frac{\partial u_x}{\partial x}\frac{\partial^2\theta}{\partial x^2} + \frac{\partial^2 u_x}{\partial x^2}\frac{\partial\theta}{\partial x} + \frac{3}{2}r^2\frac{\partial^2\theta}{\partial x^2}\left(\frac{\partial\theta}{\partial x}\right)^2\right]
$$

$$
- I_pr\omega\left[2\frac{\partial^2\theta}{\partial x\partial t}\frac{\partial\theta}{\partial x} + \frac{\partial\theta}{\partial t}\frac{\partial^2\theta}{\partial x^2}\right] + m_pg\sin\theta + m_pr\frac{\partial^2\theta}{\partial t^2} = 0 \tag{3.102}
$$

Summing up, the final problem consists of four equations to determine four unknowns: Eqs. 3.66, 3.81, 3.101 and 3.102 which relate F_x, u_x, N and θ.

3.1.2 Model for Tripping Out

For the problem of tripping out, the equations previously presented are severely simplified. This happens because the column does not suffer buckling and thus remains in contact with the lowest portion of the well for its whole length and for the whole time. During tripping out, the point C_0 from Fig. 2.11a displaces itself from (x, 0, -r) to $(x + u_x, 0, -r)$. On this case, the position vector $\vec{r}$ is given by:

$$
\vec{r}(x, t) = (x + u_x)\hat{i} - r\hat{k} \tag{3.103}
$$

The procedure now is similar than before, but some steps are no longer needed. Firstly, the unitary vector $\vec{\tau}$ from the tangential direction:

$$
\vec{\tau}(x, t) = \frac{1}{c}\frac{\partial\vec{r}}{\partial x} \tag{3.104}
$$

where c is the vector norm of $\partial \vec{r}/\partial x$. Substituting Eq. 3.103 into Eq. 3.104:

$$\vec{\tau}(x, t) = \frac{1}{c} \frac{\partial}{\partial x} \left[(x + u_x)\hat{i} - r\hat{k} \right] \tag{3.105}$$

$$\vec{\tau}(x, t) = \hat{i} \tag{3.106}$$

The vector norm c is 1 in this case and just like before, the term $\partial u_x/dx$ can be neglected. The velocity vector will be given by:

$$\vec{v}(x, t) = \frac{\partial \vec{r}}{\partial t} \tag{3.107}$$

Substituting Eq. 3.103 into Eq. 3.107:

$$\vec{v}(x, t) = \frac{\partial}{\partial t} \left[(x + u_x)\hat{i} - r\hat{k} \right] \tag{3.108}$$

$$\vec{v}(x, t) = \frac{\partial u_x}{\partial t} \hat{i} \tag{3.109}$$

Once again remembering that $\partial x/\partial t = 0$ since x is an independent coordinate. The acceleration vector $\vec{a}$ will be given by:

$$\vec{a}(x, t) = \frac{\partial \vec{v}}{\partial t} \tag{3.110}$$

Substituting Eq. 3.103 into Eq. 3.110:

$$\vec{a}(x, t) = \frac{\partial}{\partial t} \left[\frac{\partial u_x}{\partial t} \hat{i} \right] \tag{3.111}$$

$$\vec{a}(x, t) = \frac{\partial^2 u_x}{\partial t^2} \hat{i} \tag{3.112}$$

The linear momentum $\vec{P}$ of an infinitesimal element dx will be given by:

$$\vec{P}(x, t) = m_p \vec{v} dx \tag{3.113}$$

Substituting Eq. 3.109 into Eq. 3.113:

$$\vec{P}(x, t) = m_p \frac{\partial u_x}{\partial t} dx\, \hat{i} \tag{3.114}$$

The angular velocity vector $\vec{\Omega}$ will be given by:

$$\vec{\Omega}(x, t) = \omega\vec{\tau} + \frac{\partial \vec{\tau}}{\partial t} \tag{3.115}$$

Substituting Eq. 3.106 into Eq. 3.115:

$$\vec{\Omega}(x, t) = \omega\left[\hat{i}\right] + \frac{\partial}{\partial t}\left[\hat{i}\right] \tag{3.116}$$

$$\vec{\Omega}(x, t) = \omega\,\hat{i} \tag{3.117}$$

Lastly, the angular momentum $\vec{H}_0$ from an infinitesimal element dx:

$$\vec{H}_0(x, t) = \left(m_p\vec{r} \times \vec{v} + I_p\omega\vec{\tau}\right)dx \tag{3.118}$$

Substituting Eqs. 3.103, 3.106 and 3.109 into Eq. 3.118:

$$\vec{H}_0(x, t) = \left[m_p\left((x + u_x)\hat{i} - r\hat{k}\right) \times \left(\frac{\partial u_x}{\partial t}\hat{i}\right) + I_p\omega\hat{i}\right]dx \tag{3.119}$$

$$\vec{H}_0(x, t) = \left[I_p\omega\hat{i} - m_p r\frac{\partial u_x}{\partial t}\hat{j}\right]dx \tag{3.120}$$

During tripping out, there will not be any moments, since the column is not subjected to bending. This simplified the following equations. Since the column remains on the lowest portion of the well, there is no need for the unitary vectors $\hat{p}$ and $\hat{q}$. Therefore, the internal force $\vec{F}$ can be written on the $\hat{i}$, $\hat{j}$ and $\hat{k}$ coordinates:

$$\vec{F} = F_x\hat{i} + F_y\hat{j} + F_z\hat{k} \tag{3.121}$$

Taking the spatial derivative of Eq. 3.121:

$$\frac{\partial \vec{F}}{\partial x} = \frac{\partial F_x}{\partial x}\hat{i} + \frac{\partial F_y}{\partial x}\hat{j} + \frac{\partial F_z}{\partial x}\hat{k} \tag{3.122}$$

As said above, there are no bending moments, thus:

$$\vec{M} = 0 \tag{3.123}$$

Consequently, the spatial derivative of Eq. 3.123 will be:

$$\frac{\partial \vec{M}}{\partial x} = 0 \tag{3.124}$$

As in the previous case, the column own weight per unit of length $\vec{q}_p$ acts on the central axis and on the negative direction of the z axis. Therefore:

$$\vec{q}_p = -m_p g dx\hat{k} \tag{3.125}$$

The normal contact force has the same modulus than before, but it is now aligned to the z axis on its positive direction.

$$\vec{N} = Ndx\hat{k} \tag{3.126}$$

Adding up these two forces, the total external force per unit of length will be the same as before:

$$\vec{f} = \frac{\vec{q}_p}{dx} + \frac{\vec{N}}{dx} \tag{3.127}$$

Substituting Eqs. 3.125, 3.126 into Eq. 3.127:

$$\vec{f} = (-m_pg + N)\hat{k} \tag{3.128}$$

Once again, from the Strength of Materials, it is known that the internal forces are directly tied with displacements and strains. On this case, the total axial displacement u_x is the same as the axial displacement u_a, since there is not a displacement due to bending:

$$u_x = u_a \tag{3.129}$$

The displacement u_a from Eq. 3.129 is obtained once again from Hooke's Law:

$$F_x(x, t) = -EA\frac{\partial u_a}{\partial x} \tag{3.130}$$

Combining Eqs. 3.129, 3.130:

$$F_x(x, t) = -EA\frac{\partial u_x}{\partial x} \tag{3.131}$$

As before, applying Newton's Second Law to find the motion equation for the column, starting for the linear momentum:

$$\sum \vec{F}_i = \frac{\partial \vec{P}}{\partial t} \tag{3.132}$$

Substituting Eq. 3.113 into Eq. 3.132 and introducing the loadings defined through the convention from Fig. 3.1:

$$\vec{F} - \left(\vec{F} + \frac{\partial \vec{F}}{\partial x}dx\right) + \vec{f}dx = m_p\frac{\partial \vec{v}}{\partial t}dx \tag{3.133}$$

Simplifying Eq. 3.133:

$$\frac{\partial \overrightarrow{F}}{\partial x} - \overrightarrow{f} + m_p \frac{\partial \overrightarrow{v}}{\partial t} = 0 \tag{3.134}$$

Substituting Eqs. 3.112, 3.122 and 3.128 into Eq. 3.134:

$$\frac{\partial F_x}{\partial x}\hat{i} + \frac{\partial F_y}{\partial x}\hat{j} + \frac{\partial F_z}{\partial x}\hat{k} - \left(-m_p g + N\right)\hat{k} + m_p\left[\frac{\partial^2 u_x}{\partial t^2}\hat{i}\right] = 0 \tag{3.135}$$

$$\left(\frac{\partial F_x}{\partial x} + m_p\frac{\partial^2 u_x}{\partial t^2}\right)\hat{i} + \left(\frac{\partial F_y}{\partial x}\right)\hat{j} + \left(\frac{\partial F_z}{\partial x} + m_p g - N\right)\hat{k} = 0 \tag{3.136}$$

Separating Eq. 3.136 into its components, three motion equations are obtained:

$$\frac{\partial F_x}{\partial x} + m_p\frac{\partial^2 u_x}{\partial t^2} = 0 \tag{3.137}$$

$$\frac{\partial F_y}{\partial x} = 0 \tag{3.138}$$

$$\frac{\partial F_z}{\partial x} + m_p g - N = 0 \tag{3.139}$$

Substituting Eq. 3.131 into Eq. 3.137:

$$\frac{\partial}{\partial x}\left[-EA\frac{\partial u_x}{\partial x}\right] + m_p\frac{\partial^2 u_x}{\partial t^2} = 0 \tag{3.140}$$

Manipulating Eq. 3.140, an equation for the axial displacement is found:

$$EA\frac{\partial^2 u_x}{\partial x^2} - m_p\frac{\partial^2 u_x}{\partial t^2} = 0 \tag{3.141}$$

Applying Newton's Second Law now to the angular momentum:

$$\sum \overrightarrow{M}_i = \frac{\partial \overrightarrow{H}_0}{\partial t} \tag{3.142}$$

Substituting Eq. 3.118 into Eq. 3.142 and introducing the loadings defined previously on Fig. 3.1:

$$\overrightarrow{M} - \left(\overrightarrow{M} + \frac{\partial \overrightarrow{M}}{\partial x}dx\right) - \overrightarrow{r} \times \overrightarrow{F} = \frac{\partial}{\partial t}\left[(m_p\overrightarrow{r} \times \overrightarrow{v} + I_p\omega\overrightarrow{\tau})dx\right] \tag{3.143}$$

Simplifying Eq. 3.143 and taking the spatial derivative of $\overrightarrow{r}$:

$$\frac{\partial \overrightarrow{M}}{\partial x} + \overrightarrow{\tau} \times \overrightarrow{F} + \frac{\partial}{\partial t}\left[m_p\overrightarrow{r} \times \overrightarrow{v} + I_p\omega\overrightarrow{\tau}\right] = 0 \tag{3.144}$$

Substituting Eqs. 3.106, 3.120, 3.121 and 3.124 into Eq. 3.144:

$$\left(\hat{i}\right) \times \left(F_x\hat{i} + F_y\hat{j} + F_z\hat{k}\right) - m_p r\frac{\partial^2 u_x}{\partial t^2}\hat{j} = 0 \tag{3.145}$$

The term $I_p\omega\vec{\tau}$ disappears from the equation since $\vec{\tau} = \hat{i}$ and $\partial I_p\omega\hat{i}/\partial t = 0$. Manipulating Eq. 3.145:

$$\left(-F_z - m_p r\frac{\partial^2 u_x}{\partial t^2}\right)\hat{j} + F_y k = 0 \tag{3.146}$$

Separating Eq. 3.146 into its components, two motion equations are obtained:

$$F_z = -m_p r\frac{\partial^2 u_x}{\partial t^2} \tag{3.147}$$

$$F_y = 0 \tag{3.148}$$

Substituting Eq. 3.147 into Eq. 3.139:

$$\frac{\partial}{\partial x}\left[-m_p r\frac{\partial^2 u_x}{\partial t^2}\right] + m_p g - N = 0 \tag{3.149}$$

Isolating the normal contact force per unit of length N:

$$N(x, t) = -m_p r\frac{\partial^3 u_x}{\partial x\,\partial t^2} + m_p g \tag{3.150}$$

Summing up, the final problem is now only three equations for three unknowns: Eqs. 3.131, 3.141 and 3.150 which relate F_x, u_x and N.

3.1.3 Solution for Tripping in

Due to the complexity of the problem of tripping in, an analytical solution is not possible. Therefore, a numerical solution using the finite differences method will be used. For convenience, the four final equations of the model are repeated here:

$$EA\frac{\partial^2 u_x}{\partial x^2} - m_p\frac{\partial^2 u_x}{\partial t^2} + EAr^2\frac{\partial\theta}{\partial x}\frac{\partial^2\theta}{\partial x^2} = 0 \tag{3.151}$$

$$EIr\left[\frac{\partial^4\theta}{\partial x^4} - 6\left(\frac{\partial\theta}{\partial x}\right)^2\frac{\partial^2\theta}{\partial x^2}\right]$$

$$-EAr\left[\frac{\partial u_x}{\partial x}\frac{\partial^2\theta}{\partial x^2} + \frac{\partial^2 u_x}{\partial x^2}\frac{\partial\theta}{\partial x} + \frac{3}{2}r^2\frac{\partial^2\theta}{\partial x^2}\left(\frac{\partial\theta}{\partial x}\right)^2\right]$$

$$-I_p r\omega\left[2\frac{\partial^2\theta}{\partial x\partial t}\frac{\partial\theta}{\partial x} + \frac{\partial\theta}{\partial t}\frac{\partial^2\theta}{\partial x^2}\right] + m_p g\sin\theta + m_p r\frac{\partial^2\theta}{\partial t^2} = 0 \tag{3.152}$$

$$F_x(x,t) = -EA\frac{\partial u_x}{\partial x} - \frac{1}{2}EAr^2\left(\frac{\partial\theta}{\partial x}\right)^2 \tag{3.153}$$

$$N(x,t) = -EIr\left[\left(\frac{\partial\theta}{\partial x}\right)^4 - 3\left(\frac{\partial^2\theta}{\partial x^2}\right)^2 - 4\frac{\partial^3\theta}{\partial x^3}\frac{\partial\theta}{\partial x}\right]$$

$$-EAr\left[\frac{\partial u_x}{\partial x}\left(\frac{\partial\theta}{\partial x}\right)^2 + \frac{1}{2}r^2\left(\frac{\partial\theta}{\partial x}\right)^4\right]$$

$$+I_p r\omega\left[\frac{\partial^3\theta}{\partial x^2\partial t} - \frac{\partial\theta}{\partial t}\left(\frac{\partial\theta}{\partial x}\right)^2\right] + m_p g\cos\theta + m_p r\left(\frac{\partial\theta}{\partial t}\right)^2 \tag{3.154}$$

Equations 3.151, 3.152 relate directly the axial displacement u_x with the angular displacement θ, while Eqs. 3.153, 3.154 allow calculating the axial and normal forces if the displacements are known. Therefore, this set of equations is not a system; it is possible to find first the displacements and only then calculate the forces. Also, discretizing the time derivatives using the centered formula, the problem becomes implicit and decouples Eqs. 3.151, 3.152, since the coupling between u_x and θ only happens on the spatial derivatives. Discretizing Eqs. 3.151–3.154 and manipulating them:

$$U_{i,j+1} = 2U_{i,j} - U_{i,j-1} + \frac{EA\Delta t^2}{m_p\Delta x^2}\left(U_{i+1,j} - 2U_{i,j} + U_{i-1,j}\right)$$

$$+ \frac{EAr^2\Delta t^2}{2m_p\Delta x^3}\left(\theta_{i+1,j} - \theta_{i-1,j}\right)\left(\theta_{i+1,j} - 2\theta_{i,j} + \theta_{i-1,j}\right) \tag{3.155}$$

$$\frac{I_p r \omega (\theta_{i+1,j} - \theta_{i-1,j})}{4 \Delta x^2 \Delta t} \theta_{i+1,j+1}$$

$$+ \left[\frac{I_p r \omega (\theta_{i+1,j} - 2\theta_{i,j} + \theta_{i-1,j})}{2 \Delta x^2 \Delta t} - \frac{m_p r}{\Delta t^2} \right] \theta_{i,j+1}$$

$$- \frac{I_p r \omega (\theta_{i+1,j} - \theta_{i-1,j})}{4 \Delta x^2 \Delta t} \theta_{i-1,j+1}$$

$$= \frac{I_p r \omega (\theta_{i+1,j} - 2\theta_{i,j} + \theta_{i-1,j})}{2 \Delta x^2 \Delta t} \theta_{i,j-1}$$

$$- \frac{I_p r \omega (\theta_{i+1,j} - \theta_{i-1,j})}{4 \Delta x^2 \Delta t} \left(-\theta_{i+1,j-1} + \theta_{i-1,j-1} \right)$$

$$+ \frac{m_p r}{\Delta t^2} \left(-2\theta_{i,j} + \theta_{i,j-1} \right)$$

$$+ \frac{EIr}{\Delta x^4} \Big[\left(\theta_{i+2,j} - 4\theta_{i+1,j} + 6\theta_{i,j} - 4\theta_{i-1,j} + \theta_{i-2,j} \right)$$

$$- \frac{3}{2} \left(\theta_{i+1,j} - \theta_{i-1,j} \right)^2 \left(\theta_{i+1,j} - 2\theta_{i,j} + \theta_{i-1,j} \right) \Big]$$

$$- \frac{EAr}{2\Delta x^3} \Big[\left(U_{i+1,j} - U_{i-1,j} \right) \left(\theta_{i+1,j} - 2\theta_{i,j} + \theta_{i-1,j} \right)$$

$$+ \left(U_{i+1,j} - 2U_{i,j} + U_{i-1,j} \right) \left(\theta_{i+1,j} - \theta_{i-1,j} \right)$$

$$+ \frac{3r^2}{4\Delta x} \left(\theta_{i+1,j} - 2\theta_{i,j} + \theta_{i-1,j} \right) \left(\theta_{i+1,j} - \theta_{i-1,j} \right)^2 \Big]$$

$$+ m_p g \sin \theta_{i,j} \tag{3.156}$$

$$F_{i,j} = -\frac{EA}{2\Delta x} \left[\left(U_{i+1,j} - U_{i-1,j} \right) + \frac{r^2}{4\Delta x} \left(\theta_{i+1,j} - \theta_{i-1,j} \right)^2 \right] \tag{3.157}$$

$$N_{i,j} = -\frac{EIr}{\Delta x^4} \left[\frac{1}{16} \left(\theta_{i+1,j} - \theta_{i-1,j} \right)^4 - 3 \left(\theta_{i+1,j} - 2\theta_{i,j} + \theta_{i-1,j} \right)^2 \right.$$

$$\left. - \left(\theta_{i+2,j} - 2\theta_{i+1,j} + 2\theta_{i-1,j} - \theta_{i-2,j} \right) \left(\theta_{i+1,j} - \theta_{i-1,j} \right) \right]$$

$$- \frac{EAr}{8\Delta x^3} \left[\left(U_{i+1,j} - U_{i-1,j} \right) \left(\theta_{i+1,j} - \theta_{i-1,j} \right)^2 + \frac{r^2}{4\Delta x} \left(\theta_{i+1,j} - \theta_{i-1,j} \right)^4 \right]$$

$$+ \frac{I_p r \omega}{2\Delta x^2 \Delta t} \left[\left(\theta_{i+1,j} - 2\theta_{i,j} + \theta_{i-1,j} - \theta_{i+1,j-2} + 2\theta_{i,j-2} - \theta_{i-1,j-2} \right) \right.$$

$$\left. - \frac{1}{4} \left(\theta_{i,j} - \theta_{i,j-2} \right) \left(\theta_{i+1,j} - \theta_{i-1,j} \right)^2 \right]$$

$$+ m_p g \cos \theta_{i,j} + \frac{m_p r}{4\Delta t^2} \left(\theta_{i,j} - \theta_{i,j-2} \right)^2 \tag{3.158}$$

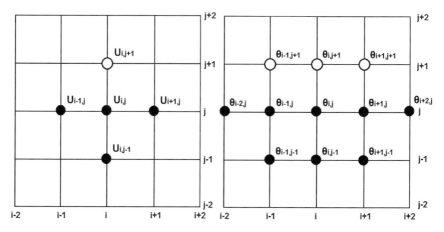

Fig. 3.2 Point mesh for variables U and θ

where subscript i denotes the space and subscript j represents time. On Eqs. 3.151, 3.152, the terms with subscript j + 1 were isolated, which represent the unknowns of the problem, as long as the values for intervals j and j–1 are known. Meanwhile, on Eqs. 3.153, 3.154, the time derivatives were discretized using the backward difference to facilitate their solutions. For the axial displacement u_x, it is easy to note that the value for each point i can be found independently of adjacent points i+1 and i–1, thus eliminating the need of solving a system. However, for the angular displacement θ, the value at each point i is dependent of the adjacent points i+1 e i–1, thus leading to a linear system of equations. It is worth pointing that despite the problem being uncoupled, the two displacements must march together in time. This happens because in order to calculate the axial displacement at interval j + 1 the angular displacement at interval j is needed and vice versa. Figure 3.2 shows the point mesh needed for solving Eqs. 3.155, 3.156. The level j + 1, marked in white, are unknowns that must be calculated, while the levels j and j–1, marked in black, represent variables already known.

The spatial discretization divides the column into N + 1 points, with points 0 and N being the extremities. Meanwhile, the time discretization starts in j = 1, with j = 1 being the initial condition for the displacement and j = 2 being the initial condition for the velocity. Therefore, the equations shown previously are valid for j = 3. The mesh for the spatial discretization can be seen on Fig. 3.3. It is important to note that besides dividing the column into N + 1 points, from i = 0 up to i = N, artificial points i = −1 and i = N + 1 must be created to discretize the boundary conditions.

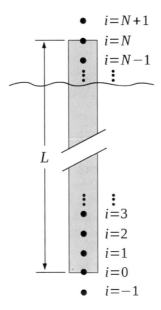

Fig. 3.3 Discretization of a column of total length L into $N+1$ points

Remains to be defined the initial conditions and the boundary conditions of the problem, so then the equations for points $i=0$ and $i=N$ and for points $j=1$ and $j=2$ can be found. Considering that the column is fixed at $x=0$ but free to move in $x=L$, the boundary conditions will be given by:

$$u(0) = 0 \tag{3.159}$$

$$\left.\frac{\partial u}{\partial x}\right|_{x=L} + \frac{r^2}{2}\left(\left.\frac{\partial \theta}{\partial x}\right|_{x=L}\right)^2 = 0 \tag{3.160}$$

$$\theta(0) = 0 \tag{3.161}$$

$$\left.\frac{\partial^2\theta}{\partial x^2}\right|_{x=0} = 0 \tag{3.162}$$

$$\left.\frac{\partial^2\theta}{\partial x^2}\right|_{x=L} = 0 \tag{3.163}$$

$$\left.\frac{\partial^3\theta}{\partial x^3}\right|_{x=L} = 0 \tag{3.164}$$

Discretizing Eqs. 3.159–3.164:

$$U_{0,j} = 0 \tag{3.165}$$

$$U_{N+1,j} - U_{N-1,j} + \frac{r^2}{4\Delta x}\left(\theta_{N+1,j} - \theta_{N-1,j}\right)^2 = 0 \rightarrow U_{N+1,j} = U_{N-1,j} - \frac{r^2}{\Delta x}\theta_{N-1,j}^2$$

$$\tag{3.166}$$

$$\theta_{0,j} = 0 \tag{3.167}$$

$$\theta_{1,j} - 2\theta_{0,j} + \theta_{-1,j} = 0 \rightarrow \theta_{-1,j} = -\theta_{1,j} \tag{3.168}$$

$$\theta_{N+1,j} - 2\theta_{N,j} + \theta_{N-1,j} = 0 \rightarrow \theta_{N+1,j} = -\theta_{N-1,j} \tag{3.169}$$

$$\theta_{N+2,j} - 2\theta_{N+1,j} + 2\theta_{N-1,j} - \theta_{N-2,j} = 0 \rightarrow \theta_{N+2,j} = \theta_{N-2,j} - 4\theta_{N-1,j} \tag{3.170}$$

From Eqs. 3.165–3.170, the equations for points i=1, i=N–1 and i=N are found substituting into Eqs. 3.155, 3.156—the point i=0 is not needed since the displacements are already known as being zero. Now, for the initial conditions, an initial displacement is imposed for u_x and θ and the initial velocities are considered zero. A small value is given to θ for convergence purposes.

$$u(x, 0) = U_0 \sin\left(\frac{\pi x}{L}\right) \tag{3.171}$$

$$\frac{\partial u(x, 0)}{\partial t} = 0 \tag{3.172}$$

$$\theta(x, 0) = 0.1 \tag{3.173}$$

$$\frac{\partial \theta(x, 0)}{\partial t} = 0 \tag{3.174}$$

Discretizing Eqs. 3.171–3.174:

$$U_{i,1} = U_0 \sin\left(\frac{\pi i \Delta x}{L}\right) \tag{3.175}$$

$$\frac{U_{i,2} - U_{i,0}}{2\Delta t} = 0 \rightarrow U_{i,0} = U_{i,2} \rightarrow U_{i,2}$$

$$= U_{i,1} + \frac{EA\Delta t^2}{2m_p\Delta x^2}\left(U_{i+1,1} - 2U_{i,1} + U_{i-1,1}\right)$$

$$+ \frac{EAr^2\Delta t^2}{4m_p\Delta x^3}\left(\theta_{i+1,1} - \theta_{i-1,1}\right)\left(\theta_{i+1,1} - 2\theta_{i,1} + \theta_{i-1,1}\right) \tag{3.176}$$

$$\theta_{i,1} = 0.1 \tag{3.177}$$

$$\frac{\theta_{i,2} - \theta_{i,0}}{2\Delta t} = 0 \rightarrow \theta_{i,0} = \theta_{i,2} \rightarrow \frac{I_p r \omega \left(\theta_{i+1,j} - \theta_{i-1,j}\right)}{2\Delta x^2 \Delta t} \theta_{i+1,2}$$

$$+ \left[\frac{I_p r \omega \left(\theta_{i+1,1} - 2\theta_{i,1} + \theta_{i-1,1}\right)}{2\Delta x^2 \Delta t} - \frac{m_p r}{\Delta t^2} \right] \theta_{i,2}$$

$$- \frac{I_p r \omega \left(\theta_{i+1,1} - \theta_{i-1,1}\right)}{2\Delta x^2 \Delta t} \theta_{i-1,2} = \frac{m_p r}{\Delta t^2} \left(-2\theta_{i,1}\right)$$

$$+ \frac{EI r}{\Delta x^4} \left[\left(\theta_{i+2,1} - 4\theta_{i+1,1} + 6\theta_{i,1} - 4\theta_{i-1,1} + \theta_{i-2,1}\right) \right.$$

$$- \frac{3}{2} \left(\theta_{i+1,1} - \theta_{i-1,1}\right)^2 \left(\theta_{i+1,1} - 2\theta_{i,1} + \theta_{i-1,1}\right) \right]$$

$$- \frac{EA r}{2\Delta x^3} \left[\left(U_{i+1,1} - U_{i-1,1}\right)\left(\theta_{i+1,1} - 2\theta_{i,1} + \theta_{i-1,1}\right) \right.$$

$$+ \left(U_{i+1,1} - 2U_{i,1} + U_{i-1,1}\right)\left(\theta_{i+1,1} - \theta_{i-1,1}\right)$$

$$+ \frac{3r^2}{4\Delta x} \left(\theta_{i+1,1} - 2\theta_{i,1} + \theta_{i-1,1}\right)\left(\theta_{i+1,1} - \theta_{i-1,1}\right)^2 \right]$$

$$+ m_p g \sin \theta_{i,1} \tag{3.178}$$

3.1.4 Solution for Tripping Out

Differently from the tripping in case, the equations for tripping out the column are simpler and possess an analytical solution. However, only the numerical solution—which will be the one used—is shown here. The three equations are repeated here for convenience:

$$EA\frac{\partial^2 u_x}{\partial x^2} - m_p \frac{\partial^2 u_x}{\partial t^2} = 0 \tag{3.179}$$

$$F_x(x, t) = -EA\frac{\partial u_x}{\partial x} \tag{3.180}$$

$$N(x, t) = -m_p r \frac{\partial^3 u_x}{\partial x \partial t^2} + m_p g \tag{3.181}$$

Discretizing Eqs. 3.179–3.181 and manipulating them:

$$U_{i,j+1} = 2U_{i,j} - U_{i,j-1} + \frac{EA\Delta t^2}{m_p \Delta x^2} \left(U_{i+1,j} - 2U_{i,j} + U_{i-1,j}\right) \tag{3.182}$$

$$F_{i,j} = -\frac{EA}{2\Delta x} \left(U_{i+1,j} - U_{i-1,j}\right) \tag{3.183}$$

$$N_{i,j} = -\frac{m_p r}{2\Delta x \Delta t^2} \left(U_{i+1,j} - U_{i-1,j} - 2U_{i+1,j-1} + 2U_{i-1,j-1} + U_{i+1,j-2} - U_{i-1,j-2}\right)$$
$$+ m_p g$$

$$(3.184)$$

As in the tripping in case, first Eq. 3.182 is solved to find the axial displacements and only then Eqs. 3.183, 3.184 are solved to find the axial and normal forces. Remains to be defined the boundary and initial conditions. As said before, the column is fixed in $x = 0$ and free on $x = L$. The boundary conditions will then be given by:

$$u(0) = 0 \qquad\qquad (3.185)$$

$$\left.\frac{\partial u}{\partial x}\right|_{x=L} = 0 \qquad\qquad (3.186)$$

Discretizing Eqs. 3.185–3.186:

$$U_{0,j} = 0 \qquad\qquad (3.187)$$

$$U_{N+1,j} - U_{N-1,j} = 0 \rightarrow U_{N+1,j} = U_{N-1,j} \qquad\qquad (3.188)$$

From Eqs. 3.187, 3.188, the equations for the points $i = 1$, $i = N-1$ and $i = N$ are found by substituting on Eq. 3.182. Lastly, the initial conditions will be the same from the tripping in case:

$$u(x, 0) = U_0 \sin\left(\frac{\pi x}{L}\right) \qquad\qquad (3.189)$$

$$\frac{\partial u(x, 0)}{\partial t} = 0 \qquad\qquad (3.190)$$

Discretizing Eqs. 3.189, 3.190:

$$U_{i,1} = U_0 \sin\left(\frac{\pi i \Delta x}{L}\right) \qquad\qquad (3.191)$$

$$\frac{U_{i,2} - U_{i,0}}{2\Delta t} = 0 \rightarrow U_{i,0} = U_{i,2} \rightarrow U_{i,2} = U_{i,1} + \frac{EA\Delta t^2}{2m_p \Delta x^2}\left(U_{i+1,1} - 2U_{i,1} + U_{i-1,1}\right)$$

$$(3.192)$$

3.2 Model II: Columns with Friction

The effect of the friction force on the column buckling problem had already been studied previously on the literature (Mitchell 1986; Mitchell 1996; Mitchell 2007; Gao and Miska 2009). However, in none of said works the friction force was considered as being part of a dynamic problem, but only for static cases. The objective here is to unify the ideas from Gao and Miska (2009)—whose model is static and has friction—with the ideas from Gao and Miska (2010)—whose model is dynamic but without friction.

The friction force, differently from other external forces such as the weight and normal contact, does not have a fixed direction as time passes. Its direction is always opposite to the direction of the velocity; since the velocity can change its direction as time passes, the direction of the friction force will change as well. Besides, since the column is free to displace angularly inside the well during its tripping in, two possible scenarios can occur: the column can roll without slipping or roll while slipping. On the first case, the friction force is static—since there is no relative motion between the column and the wellbore—and its modulus can be any value from zero up to the maximum static friction—in which case the column starts slipping. Meanwhile, on the second case, the friction force is dynamic, because there is relative motion between the column and the wellbore; therefore, the friction force has a fixed modulus and can be obtained if the dynamic friction coefficient between the two surfaces and the normal contact force is known. On this book, it will be assumed that the column rolls while slipping. This hypothesis allows writing the friction force as a function of the normal contact force, thus reducing the number of variables—if the column could roll without slipping, the friction force would be an extra variable, since it would not be written as a function of the normal contact force.

Finally, remains to be defined the direction of the friction force. Both during tripping in and tripping out, the friction force will possess a component on $\hat{i}$, whose sign will depend on the direction of the axial velocity $\partial u_x / \partial t$ . . However, during tripping in, the column also suffers an angular displacement, which will result into a lateral friction on the $\hat{q}$ direction, as shown on Fig. 3.4, whose sign will depend on the angular velocity $\partial\theta / \partial t$; this component is not present during tripping out.

Based on these hypotheses, it is possible to characterize the friction force and then repeat the procedure from the previous section to obtain new equations of motion for the problem. This will be done for both tripping in and tripping out.

3.2.1 Model for Tripping in

The friction force has two components and thus can be written as:

$$\overrightarrow{F}_f(x, t) = \overrightarrow{F}_{f1}(x, t)\hat{i} + \overrightarrow{F}_{f2}(x, t)\hat{q} \qquad (3.193)$$

$\overrightarrow{F}_{f1}$ is the axial component while $\overrightarrow{F}_{f2}$ is the lateral one. The modulus of component $\overrightarrow{F}_{f1}$ will be given by:

$$\left|\overrightarrow{F}_{f1}\right| = f_1 N dx \qquad (3.194)$$

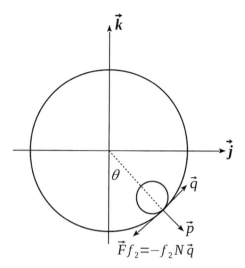

Fig. 3.4 Lateral friction caused by the column angular motion inside the well during its tripping in

where f_1 is the dynamic friction coefficient on the axial direction, N is the normal contact force per unit of length and dx is an infinitesimal element of length. Since $\vec{F}_{f1}$ is dependent of the sign of $\partial u_x / \partial t$:

$$\vec{F}_{f1}(x, t) = -\text{sgn}\left(\frac{\partial u_x}{\partial t}\right) f_1 N dx \,\hat{i} \tag{3.195}$$

where sgn is the sign function. If the velocity is positive, the sign function has value of 1 and the friction force is on the negative direction; if the velocity is negative, the sign function has value of -1 and the friction force is on the positive direction; and if the velocity is zero, the sign function has value 0 and then there is no dynamic friction force. The same procedure is valid for the component $\vec{F}_{f2}$, but on this case it is dependent of the sign of $\partial\theta / \partial t$:

$$\left|\vec{F}_{f2}\right| = f_2 N dx \tag{3.196}$$

$$\vec{F}_{f2}(x, t) = -\text{sgn}\left(\frac{\partial\theta}{\partial t}\right) f_2 N dx \hat{q} \tag{3.197}$$

where f_2 is the dynamic friction coefficient on the lateral direction. It is possible to manipulate the two friction coefficients f_1 and f_2 by introducing a single total dynamic friction coefficient f. Through Eq. 3.193, the following relation is valid:

$$\left|\vec{F}_f\right|^2 = \left|\vec{F}_{f1}\right|^2 + \left|\vec{F}_{f2}\right|^2 \tag{3.198}$$

Substituting the modulus of the vectors:

$$f^2 N^2 dx^2 = f_1^2 N^2 dx^2 + f_2^2 N^2 dx^2 \tag{3.199}$$

$$f^2 = f_1^2 + f_2^2 \tag{3.200}$$

Another possible relation is to consider that f_1 and f_2 are proportional to the respective axial velocity v_1 and lateral velocity v_2. For small velocities, this linear relation is acceptable. Therefore:

$$\frac{f_1}{f_2} = \frac{v_1(x, t)}{v_2(x, t)} \tag{3.201}$$

where v_1 and v_2 are the components of the velocity on the $\hat{i}$ and $\hat{q}$ directions, as given by Eq. 3.27. Starting from Eq. 3.201, each friction coefficient can be isolated:

$$f_1 = f_2 \frac{v_1(x, t)}{v_2(x, t)} \tag{3.202}$$

$$f_2 = f_1 \frac{v_2(x, t)}{v_1(x, t)} \tag{3.203}$$

Combining Eq. 3.203 with Eq. 3.200:

$$f^2 = f_1^2 + f_1^2 \frac{v_2^2}{v_1^2} \tag{3.204}$$

$$v_1^2 f^2 = v_1^2 f_1^2 + v_2^2 f_1^2 \tag{3.205}$$

$$f_1^2 = \frac{v_1^2 f^2}{v_1^2 + v_2^2} \tag{3.206}$$

$$f_1 = \frac{|v_1| f}{\sqrt{v_1^2 + v_2^2}} \tag{3.207}$$

Now combining Eq. 3.202 with Eq. 3.200:

$$f^2 = f_2^2 \frac{v_1^2}{v_2^2} + f_2^2 \tag{3.208}$$

$$v_2^2 f^2 = v_1^2 f_2^2 + v_2^2 f_2^2 \tag{3.209}$$

$$f_2^2 = \frac{v_2^2 f^2}{v_1^2 + v_2^2} \tag{3.210}$$

$$f_2 = \frac{|v_2| f}{\sqrt{v_1^2 + v_2^2}} \tag{3.211}$$

Substituting Eqs. 3.195, 3.197, 3.207 and 3.211 into Eq. 3.193:

$$\vec{F}_f(x, t) = -\text{sgn}\left(\frac{\partial u_x}{\partial t}\right)\frac{|v_1|f}{\sqrt{v_1^2 + v_2^2}}N dx\,\hat{i} - \text{sgn}\left(\frac{\partial \theta}{\partial t}\right)\frac{|v_2|f}{\sqrt{v_1^2 + v_2^2}}N dx\,\hat{q} \quad (3.212)$$

$$\vec{F}_f(x, t) = -\frac{f N dx}{\sqrt{v_1^2 + v_2^2}}\left[\text{sgn}\left(\frac{\partial u_x}{\partial t}\right)|v_1|\hat{i} + \text{sgn}\left(\frac{\partial \theta}{\partial t}\right)|v_2|\hat{q}\right] \quad (3.213)$$

Knowing that $v_1 = \partial u_x/\partial t$ and $v_2 = r\,\partial\theta/\partial t$, according to Eq. 3.27, and substituting into Eq. 3.213:

$$\vec{F}_f(x, t) = -\frac{f N dx}{\sqrt{\left(\frac{\partial u_x}{\partial t}\right)^2 + r^2\left(\frac{\partial \theta}{\partial t}\right)^2}}\left[\text{sgn}\left(\frac{\partial u_x}{\partial t}\right)\left|\frac{\partial u_x}{\partial t}\right|\hat{i} + \text{sgn}\left(\frac{\partial \theta}{\partial t}\right)r\left|\frac{\partial \theta}{\partial t}\right|\hat{q}\right] \quad (3.214)$$

Therefore, the total external force per unit of length is now given by:

$$\vec{f} = \frac{\vec{q}_p}{dx} + \frac{\vec{N}}{dx} + \frac{\vec{F}_f}{dx} \quad (3.215)$$

Substituting Eqs. 3.57, 3.58 and 3.214 into Eq. 3.215:

$$\vec{f} = m_p g\cos\theta\,\hat{p} - m_p g\sin\theta\,\hat{q} - N\hat{p}$$
$$-\frac{f N}{\sqrt{\left(\frac{\partial u_x}{\partial t}\right)^2 + r^2\left(\frac{\partial \theta}{\partial t}\right)^2}}\left[\text{sgn}\left(\frac{\partial u_x}{\partial t}\right)\left|\frac{\partial u_x}{\partial t}\right|\hat{i} + \text{sgn}\left(\frac{\partial \theta}{\partial t}\right)r\left|\frac{\partial \theta}{\partial t}\right|\hat{q}\right] \quad (3.216)$$

$$\vec{f} = -\frac{\text{sgn}\left(\frac{\partial u_x}{\partial t}\right)f N\left|\frac{\partial u_x}{\partial t}\right|}{\sqrt{\left(\frac{\partial u_x}{\partial t}\right)^2 + r^2\left(\frac{\partial \theta}{\partial t}\right)^2}}\hat{i} + \left(m_p g\cos\theta - N\right)\hat{p}$$
$$-\left(m_p g\sin\theta + \frac{\text{sgn}\left(\frac{\partial \theta}{\partial t}\right)f N r\left|\frac{\partial \theta}{\partial t}\right|}{\sqrt{\left(\frac{\partial u_x}{\partial t}\right)^2 + r^2\left(\frac{\partial \theta}{\partial t}\right)^2}}\right)\hat{q} \quad (3.217)$$

The summation of forces is still given by Eq. 3.74:

$$\frac{\partial \vec{F}}{\partial x} - \vec{f} + m_p\frac{\partial \vec{v}}{\partial t} = 0 \quad (3.218)$$

Substituting Eqs. 3.31, 3.49 and 3.217 into Eq. 3.218:

$$\frac{\partial F_x}{\partial x}\hat{i} + \left(\frac{\partial F_r}{\partial x} - F_\theta\frac{\partial\theta}{\partial x}\right)\hat{p} + \left(\frac{\partial F_\theta}{\partial x} + F_r\frac{\partial\theta}{\partial x}\right)\hat{q}$$

$$+ \frac{\text{sgn}\left(\frac{\partial u_x}{\partial t}\right)fN\left|\frac{\partial u_x}{\partial t}\right|}{\sqrt{\left(\frac{\partial u_x}{\partial t}\right)^2 + r^2\left(\frac{\partial\theta}{\partial t}\right)^2}}\hat{i} - \left(m_pg\cos\theta - N\right)\hat{p}$$

$$+ \left(m_pg\sin\theta + \frac{\text{sgn}\left(\frac{\partial\theta}{\partial t}\right)fNr\left|\frac{\partial\theta}{\partial t}\right|}{\sqrt{\left(\frac{\partial u_x}{\partial t}\right)^2 + r^2\left(\frac{\partial\theta}{\partial t}\right)^2}}\right)\hat{q}$$

$$+ m_p\left[\frac{\partial^2 u_x}{\partial t^2}\hat{i} - r\left(\frac{\partial\theta}{\partial t}\right)^2\hat{p} + r\frac{\partial^2\theta}{\partial t^2}\hat{q}\right] = 0 \tag{3.219}$$

$$\left[\frac{\partial F_x}{\partial x} + \frac{\text{sgn}\left(\frac{\partial u_x}{\partial t}\right)fN\left|\frac{\partial u_x}{\partial t}\right|}{\sqrt{\left(\frac{\partial u_x}{\partial t}\right)^2 + r^2\left(\frac{\partial\theta}{\partial t}\right)^2}} + m_p\frac{\partial^2 u_x}{\partial t^2}\right]\hat{i}$$

$$+ \left[\frac{\partial F_r}{\partial x} - F_\theta\frac{\partial\theta}{\partial x} + N - m_pg\cos\theta - m_pr\left(\frac{\partial\theta}{\partial t}\right)^2\right]\hat{p}$$

$$+ \left[\frac{\partial F_\theta}{\partial x} + F_r\frac{\partial\theta}{\partial x} + m_pg\sin\theta + \frac{\text{sgn}\left(\frac{\partial\theta}{\partial t}\right)fNr\left|\frac{\partial\theta}{\partial t}\right|}{\sqrt{\left(\frac{\partial u_x}{\partial t}\right)^2 + r^2\left(\frac{\partial\theta}{\partial t}\right)^2}} + m_pr\frac{\partial^2\theta}{\partial t^2}\right]\hat{q}$$

$$= 0 \tag{3.220}$$

Separating Eq. 3.220 into components:

$$\frac{\partial F_x}{\partial x} + \frac{\text{sgn}\left(\frac{\partial u_x}{\partial t}\right)fN\left|\frac{\partial u_x}{\partial t}\right|}{\sqrt{\left(\frac{\partial u_x}{\partial t}\right)^2 + r^2\left(\frac{\partial\theta}{\partial t}\right)^2}} + m_p\frac{\partial^2 u_x}{\partial t^2} = 0 \tag{3.221}$$

$$\frac{\partial F_r}{\partial x} - F_\theta\frac{\partial\theta}{\partial x} + N - m_pg\cos\theta - m_pr\left(\frac{\partial\theta}{\partial t}\right)^2 = 0 \tag{3.222}$$

$$\frac{\partial F_\theta}{\partial x} + F_r\frac{\partial\theta}{\partial x} + m_pg\sin\theta + \frac{\text{sgn}\left(\frac{\partial\theta}{\partial t}\right)fNr\left|\frac{\partial\theta}{\partial t}\right|}{\sqrt{\left(\frac{\partial u_x}{\partial t}\right)^2 + r^2\left(\frac{\partial\theta}{\partial t}\right)^2}} + m_pr\frac{\partial^2\theta}{\partial t^2} = 0 \tag{3.223}$$

Looking at the analysis made on the previous section, it is easy to conclude that the calculations to find the components F_x, F_r and F_θ from $\overrightarrow{F}$ will not change with the introduction of the friction force. Therefore, substituting Eqs. 3.66, 3.92 and 3.95 into Eqs. 3.221–3.223, with Eqs. 3.37, 3.38:

$$\frac{\partial}{\partial x}\left[-EA\frac{\partial u_x}{\partial x} - \frac{1}{2}EAr^2\left(\frac{\partial\theta}{\partial x}\right)^2\right] + \frac{\text{sgn}\left(\frac{\partial u_x}{\partial t}\right)fN\left|\frac{\partial u_x}{\partial t}\right|}{\sqrt{\left(\frac{\partial u_x}{\partial t}\right)^2 + r^2\left(\frac{\partial\theta}{\partial t}\right)^2}} + m_p\frac{\partial^2 u_x}{\partial t^2} = 0 \tag{3.224}$$

$$\frac{\partial}{\partial x}\left[-3EIr\frac{\partial\theta}{\partial x}\frac{\partial^2\theta}{\partial x^2} - I_p\omega r\frac{\partial^2\theta}{\partial x\partial t}\right] - \left[EIr\left[\frac{\partial^3\theta}{\partial x^3} - \left(\frac{\partial\theta}{\partial x}\right)^3\right]\right.$$

$$+\left[-EA\frac{\partial u_x}{\partial x} - \frac{1}{2}EAr^2\left(\frac{\partial\theta}{\partial x}\right)^2\right]r\frac{\partial\theta}{\partial x}$$

$$+I_p\omega\left(-r\frac{\partial\theta}{\partial t}\frac{\partial\theta}{\partial x}\right)\right]\frac{\partial\theta}{\partial x} + N - m_pg\cos\theta - m_pr\left(\frac{\partial\theta}{\partial t}\right)^2 = 0 \quad (3.225)$$

$$\frac{\partial}{\partial x}\left[EIr\left[\frac{\partial^3\theta}{\partial x^3} - \left(\frac{\partial\theta}{\partial x}\right)^3\right] + \left[-EA\frac{\partial u_x}{\partial x} - \frac{1}{2}EAr^2\left(\frac{\partial\theta}{\partial x}\right)^2\right]r\frac{\partial\theta}{\partial x}\right.$$

$$+I_p\omega\left(-r\frac{\partial\theta}{\partial t}\frac{\partial\theta}{\partial x}\right)\right] + \left[-3EIr\frac{\partial\theta}{\partial x}\frac{\partial^2\theta}{\partial x^2} - I_p\omega r\frac{\partial^2\theta}{\partial x\partial t}\right]\frac{\partial\theta}{\partial x}$$

$$+ m_pg\sin\theta + \frac{\text{sgn}\left(\frac{\partial\theta}{\partial t}\right)fNr\left|\frac{\partial\theta}{\partial t}\right|}{\sqrt{\left(\frac{\partial u_x}{\partial t}\right)^2 + r^2\left(\frac{\partial\theta}{\partial t}\right)^2}} + m_pr\frac{\partial^2\theta}{\partial t^2} = 0 \quad (3.226)$$

Manipulating Eqs. 3.224–3.226:

$$EA\frac{\partial^2 u_x}{\partial x^2} - m_p\frac{\partial^2 u_x}{\partial t^2} + EAr^2\frac{\partial\theta}{\partial x}\frac{\partial^2\theta}{\partial x^2} - \frac{\text{sgn}\left(\frac{\partial u_x}{\partial t}\right)fN\left|\frac{\partial u_x}{\partial t}\right|}{\sqrt{\left(\frac{\partial u_x}{\partial t}\right)^2 + r^2\left(\frac{\partial\theta}{\partial t}\right)^2}} = 0 \quad (3.227)$$

$$N = -EIr\left[\left(\frac{\partial\theta}{\partial x}\right)^4 - 3\left(\frac{\partial^2\theta}{\partial x^2}\right)^2 - 4\frac{\partial^3\theta}{\partial x^3}\frac{\partial\theta}{\partial x}\right] - EAr\left[\frac{\partial u_x}{\partial x}\left(\frac{\partial\theta}{\partial x}\right)^2 + \frac{1}{2}r^2\left(\frac{\partial\theta}{\partial x}\right)^4\right]$$

$$+ I_p r\omega\left[\frac{\partial^3\theta}{\partial x^2\partial t} - \frac{\partial\theta}{\partial t}\left(\frac{\partial\theta}{\partial x}\right)^2\right] + m_pg\cos\theta + m_pr\left(\frac{\partial\theta}{\partial t}\right)^2 \quad (3.228)$$

$$EIr\left[\frac{\partial^4\theta}{\partial x^4} - 6\left(\frac{\partial\theta}{\partial x}\right)^2\frac{\partial^2\theta}{\partial x^2}\right] - EAr\left[\frac{\partial u_x}{\partial x}\frac{\partial^2\theta}{\partial x^2} + \frac{\partial^2 u_x}{\partial x^2}\frac{\partial\theta}{\partial x} + \frac{3}{2}r^2\frac{\partial^2\theta}{\partial x^2}\left(\frac{\partial\theta}{\partial x}\right)^2\right]$$

$$- I_p r\omega\left[2\frac{\partial^2\theta}{\partial x\partial t}\frac{\partial\theta}{\partial x} + \frac{\partial\theta}{\partial t}\frac{\partial^2\theta}{\partial x^2}\right] + m_pg\sin\theta + m_pr\frac{\partial^2\theta}{\partial t^2} + \frac{\text{sgn}\left(\frac{\partial\theta}{\partial t}\right)fNr\left|\frac{\partial\theta}{\partial t}\right|}{\sqrt{\left(\frac{\partial u_x}{\partial t}\right)^2 + r^2\left(\frac{\partial\theta}{\partial t}\right)^2}} = 0$$

$$(3.229)$$

Equation 3.228 for the normal contact force is exactly the same as Eq. 3.101 obtained previously without friction. Substituting Eq. 3.228 into Eqs. 3.227 and 3.229:

$$EA\frac{\partial^2 u_x}{\partial x^2} - m_p\frac{\partial^2 u_x}{\partial t^2} + EAr^2\frac{\partial\theta}{\partial x}\frac{\partial^2\theta}{\partial x^2}$$

$$-\frac{sgn\left(\frac{\partial u_x}{\partial t}\right)f\left|\frac{\partial u_x}{\partial t}\right|}{\sqrt{\left(\frac{\partial u_x}{\partial t}\right)^2 + r^2\left(\frac{\partial\theta}{\partial t}\right)^2}}\left\{-EIr\left[\left(\frac{\partial\theta}{\partial x}\right)^4 - 3\left(\frac{\partial^2\theta}{\partial x^2}\right)^2 - 4\frac{\partial^3\theta}{\partial x^3}\frac{\partial\theta}{\partial x}\right]\right.$$

$$-EAr\left[\frac{\partial u_x}{\partial x}\left(\frac{\partial\theta}{\partial x}\right)^2 + \frac{1}{2}r^2\left(\frac{\partial\theta}{\partial x}\right)^4\right] + I_p r\omega\left[\frac{\partial^3\theta}{\partial x^2\partial t} - \frac{\partial\theta}{\partial t}\left(\frac{\partial\theta}{\partial x}\right)^2\right]$$

$$\left. + m_p g\cos\theta + m_p r\left(\frac{\partial\theta}{\partial t}\right)^2\right\} = 0 \tag{3.230}$$

$$EIr\left[\frac{\partial^4\theta}{\partial x^4} - 6\left(\frac{\partial\theta}{\partial x}\right)^2\frac{\partial^2\theta}{\partial x^2}\right] - EAr\left[\frac{\partial u_x}{\partial x}\frac{\partial^2\theta}{\partial x^2} + \frac{\partial^2 u_x}{\partial x^2}\frac{\partial\theta}{\partial x} + \frac{3}{2}r^2\frac{\partial^2\theta}{\partial x^2}\left(\frac{\partial\theta}{\partial x}\right)^2\right]$$

$$-I_p r\omega\left[2\frac{\partial^2\theta}{\partial x\partial t}\frac{\partial\theta}{\partial x} + \frac{\partial\theta}{\partial t}\frac{\partial^2\theta}{\partial x^2}\right] + m_p g\sin\theta + m_p r\frac{\partial^2\theta}{\partial t^2}$$

$$+\frac{sgn\left(\frac{\partial\theta}{\partial t}\right)fr\left|\frac{\partial\theta}{\partial t}\right|}{\sqrt{\left(\frac{\partial u_x}{\partial t}\right)^2 + r^2\left(\frac{\partial\theta}{\partial t}\right)^2}}\left\{-EIr\left[\left(\frac{\partial\theta}{\partial x}\right)^4 - 3\left(\frac{\partial^2\theta}{\partial x^2}\right)^2 - 4\frac{\partial^3\theta}{\partial x^3}\frac{\partial\theta}{\partial x}\right]\right.$$

$$-EAr\left[\frac{\partial u_x}{\partial x}\left(\frac{\partial\theta}{\partial x}\right)^2 + \frac{1}{2}r^2\left(\frac{\partial\theta}{\partial x}\right)^4\right] + I_p r\omega\left[\frac{\partial^3\theta}{\partial x^2\partial t} - \frac{\partial\theta}{\partial t}\left(\frac{\partial\theta}{\partial x}\right)^2\right]$$

$$\left. + m_p g\cos\theta + m_p r\left(\frac{\partial\theta}{\partial t}\right)^2\right\} = 0 \tag{3.231}$$

Equations 3.230, 3.231 allow calculating the axial displacement u_x and angular displacement θ. Knowing the displacements, the forces F_x and N can then be calculated as well.

3.2.2 Model for Tripping Out

As in the model without friction, the motion equations become simplified for the tripping out problem. The friction force, in this case, has only one component and is given by:

$$\vec{F}_f(x, t) = \vec{F}_{fl}(x, t)\hat{i} \tag{3.232}$$

Similarly, this component will be given by:

$$\left|\vec{F}_{fl}\right| = f_l N dx \tag{3.233}$$

$$\vec{F}_{f1}(x, t) = -\text{sgn}\left(\frac{\partial u_x}{\partial t}\right) f_1 N dx\, \hat{i} \tag{3.234}$$

However, in this case, since there is not another friction component, $f = f_1$. Substituting Eq. 3.234 into Eq. 3.232:

$$\vec{F}_f(x, t) = -\text{sgn}\left(\frac{\partial u_x}{\partial t}\right) f N dx\, \hat{i} \tag{3.235}$$

It is interesting to point that the $\hat{i}$ component for the friction force during tripping out, given by Eq. 3.235 is different from the $\hat{i}$ component for the friction force during tripping in. This suggests that the friction force in the axial direction is, indeed, different during tripping in and tripping out the column. The total external force per unit of length will be given by:

$$\vec{F} = \frac{\vec{q}_p}{dx} + \frac{\vec{N}}{dx} + \frac{\vec{F}_f}{dx} \tag{3.236}$$

Substituting Eqs. 3.125, 3.126 and 3.235 into Eq. 3.236:

$$\vec{f} = -m_p g \hat{k} + N \hat{k} - \text{sgn}\left(\frac{\partial u_x}{\partial t}\right) f N \hat{i} \tag{3.237}$$

$$\vec{f} = -\text{sgn}\left(\frac{\partial u_x}{\partial t}\right) f N \hat{i} + \left(-m_p g + N\right) \hat{k} \tag{3.238}$$

The summation of forces will be given by Eq. 3.134. Substituting Eqs. 3.112, 3.122 and 3.238 into Eq. 3.134:

$$\frac{\partial F_x}{\partial x}\hat{i} + \frac{\partial F_y}{\partial x}\hat{j} + \frac{\partial F_z}{\partial x}\hat{k} + \text{sgn}\left(\frac{\partial u_x}{\partial t}\right) f N \hat{i} - \left(-m_p g + N\right)\hat{k} + m_p\left[\frac{\partial^2 u_x}{\partial t^2}\hat{i}\right] = 0 \tag{3.239}$$

$$\left(\frac{\partial F_x}{\partial x} + m_p\frac{\partial^2 u_x}{\partial t^2} + \text{sgn}\left(\frac{\partial u_x}{\partial t}\right) f N\right)\hat{i} + \left(\frac{\partial F_y}{\partial x}\right)\hat{j} + \left(\frac{\partial F_z}{\partial x} + m_p g - N\right)\hat{k} = 0 \tag{3.240}$$

Separating Eq. 3.240 into components:

$$\frac{\partial F_x}{\partial x} + m_p\frac{\partial^2 u_x}{\partial t^2} + \text{sgn}\left(\frac{\partial u_x}{\partial t}\right) f N = 0 \tag{3.241}$$

$$\frac{\partial F_y}{\partial x} = 0 \tag{3.242}$$

$$\frac{\partial F_z}{\partial x} + m_p g - N = 0 \tag{3.243}$$

Once again looking at the previous analysis, it is easy to see that the calculations for F_x, F_y and F_z of $\overrightarrow{F}$ will not change with the friction force. Therefore, substituting Eqs. 3.131, 3.147 and 3.148 into Eqs. 3.241–3.243, it is easy to note that Eq. 3.242 becomes irrelevant and only two equations remain:

$$\frac{\partial}{\partial x}\left[-EA\frac{\partial u_x}{\partial x}\right] + m_p\frac{\partial^2 u_x}{\partial t^2} + \text{sgn}\left(\frac{\partial u_x}{\partial t}\right)fN = 0 \qquad (3.244)$$

$$\frac{\partial}{\partial x}\left[-m_p r\frac{\partial^2 u_x}{\partial t^2}\right] + m_p g - N = 0 \qquad (3.245)$$

Manipulating Eqs. 3.244–3.245:

$$EA\frac{\partial^2 u_x}{\partial x^2} - m_p\frac{\partial^2 u_x}{\partial t^2} - \text{sgn}\left(\frac{\partial u_x}{\partial t}\right)fN = 0 \qquad (3.246)$$

$$N = -m_p r\frac{\partial^3 u_x}{\partial x\partial t^2} + m_p g \qquad (3.247)$$

Substituting Eq. 3.247 into Eq. 3.246:

$$EA\frac{\partial^2 u_x}{\partial x^2} - m_p\frac{\partial^2 u_x}{\partial t^2} - \text{sgn}\left(\frac{\partial u_x}{\partial t}\right)f\left[-m_p r\frac{\partial^3 u_x}{\partial x\partial t^2} + m_p g\right] = 0 \qquad (3.248)$$

Equation 3.248 allows calculating the axial displacement u_x. After that, the forces F_x and N can be calculated.

3.2.3 Solution for Tripping in

As in the previous case, the solution here must be numeric. Repeating the four final equations:

$$
\begin{aligned}
&EA\frac{\partial^2 u_x}{\partial x^2} - m_p\frac{\partial^2 u_x}{\partial t^2} + EAr^2\frac{\partial\theta}{\partial x}\frac{\partial^2\theta}{\partial x^2}\\
&- \frac{\text{sgn}\left(\frac{\partial u_x}{\partial t}\right)f\left|\frac{\partial u_x}{\partial t}\right|}{\sqrt{\left(\frac{\partial u_x}{\partial t}\right)^2 + r^2\left(\frac{\partial\theta}{\partial t}\right)^2}}\left\{-EIr\left[\left(\frac{\partial\theta}{\partial x}\right)^4 - 3\left(\frac{\partial^2\theta}{\partial x^2}\right)^2 - 4\frac{\partial^3\theta}{\partial x^3}\frac{\partial\theta}{\partial x}\right]\right.\\
&- EAr\left[\frac{\partial u_x}{\partial x}\left(\frac{\partial\theta}{\partial x}\right)^2 + \frac{1}{2}r^2\left(\frac{\partial\theta}{\partial x}\right)^4\right] + I_p r\omega\left[\frac{\partial^3\theta}{\partial x^2\partial t} - \frac{\partial\theta}{\partial t}\left(\frac{\partial\theta}{\partial x}\right)^2\right]\\
&\left. + m_p g\cos\theta + m_p r\left(\frac{\partial\theta}{\partial t}\right)^2\right\} = 0 \qquad (3.249)
\end{aligned}
$$

$$
EIr\left[\frac{\partial^4\theta}{\partial x^4} - 6\left(\frac{\partial\theta}{\partial x}\right)^2\frac{\partial^2\theta}{\partial x^2}\right] - EAr\left[\frac{\partial u_x}{\partial x}\frac{\partial^2\theta}{\partial x^2} + \frac{\partial^2 u_x}{\partial x^2}\frac{\partial\theta}{\partial x} + \frac{3}{2}r^2\frac{\partial^2\theta}{\partial x^2}\left(\frac{\partial\theta}{\partial x}\right)^2\right]
$$

$$
- I_p r\omega\left[2\frac{\partial^2\theta}{\partial x\partial t}\frac{\partial\theta}{\partial x} + \frac{\partial\theta}{\partial t}\frac{\partial^2\theta}{\partial x^2}\right] + m_p g\sin\theta + m_p r\frac{\partial^2\theta}{\partial t^2}
$$

$$
+ \frac{\mathrm{sgn}\left(\frac{\partial\theta}{\partial t}\right)fr\left|\frac{\partial\theta}{\partial t}\right|}{\sqrt{\left(\frac{\partial u_x}{\partial t}\right)^2 + r^2\left(\frac{\partial\theta}{\partial t}\right)^2}}\left\{-EIr\left[\left(\frac{\partial\theta}{\partial x}\right)^4 - 3\left(\frac{\partial^2\theta}{\partial x^2}\right)^2 - 4\frac{\partial^3\theta}{\partial x^3}\frac{\partial\theta}{\partial x}\right]\right.
$$

$$
- EAr\left[\frac{\partial u_x}{\partial x}\left(\frac{\partial\theta}{\partial x}\right)^2 + \frac{1}{2}r^2\left(\frac{\partial\theta}{\partial x}\right)^4\right] + I_p r\omega\left[\frac{\partial^3\theta}{\partial x^2\partial t} - \frac{\partial\theta}{\partial t}\left(\frac{\partial\theta}{\partial x}\right)^2\right]
$$

$$
\left. + m_p g\cos\theta + m_p r\left(\frac{\partial\theta}{\partial t}\right)^2\right\} = 0 \tag{3.250}
$$

$$
F_x = -EA\frac{\partial u_x}{\partial x} - \frac{1}{2}EAr^2\left(\frac{\partial\theta}{\partial x}\right)^2 \tag{3.251}
$$

$$
N = -EIr\left[\left(\frac{\partial\theta}{\partial x}\right)^4 - 3\left(\frac{\partial^2\theta}{\partial x^2}\right)^2 - 4\frac{\partial^3\theta}{\partial x^3}\frac{\partial\theta}{\partial x}\right] - EAr\left[\frac{\partial u_x}{\partial x}\left(\frac{\partial\theta}{\partial x}\right)^2 + \frac{1}{2}r^2\left(\frac{\partial\theta}{\partial x}\right)^4\right]
$$

$$
+ I_p r\omega\left[\frac{\partial^3\theta}{\partial x^2\partial t} - \frac{\partial\theta}{\partial t}\left(\frac{\partial\theta}{\partial x}\right)^2\right] + m_p g\cos\theta + m_p r\left(\frac{\partial\theta}{\partial t}\right)^2 \tag{3.252}
$$

Discretizing Eqs. 3.249–3.252 and manipulating them:

$$U_{i,j+1}$$

$$= 2U_{i,j} - U_{i,j-1} + \frac{EA\Delta t^2}{m_p \Delta x^2}\left(U_{i+1,j} - 2U_{i,j} + U_{i-1,j}\right)$$

$$+ \frac{EAr^2\Delta t^2}{2m_p\Delta x^3}\left(\theta_{i+1,j} - \theta_{i-1,j}\right)\left(\theta_{i+1,j} - 2\theta_{i,j} + \theta_{i-1,j}\right)$$

$$- \frac{\text{sgn}\left(U_{i,j} - U_{i,j-1}\right)f\Delta t\left|U_{i,j} - U_{i,j-2}\right|}{2m_p\sqrt{\left(\frac{U_{i,j}-U_{i,j-2}}{2\Delta t}\right)^2 + r^2\left(\frac{\theta_{i,j}-\theta_{i,j-2}}{2\Delta t}\right)^2}}$$

$$\left\{-\frac{EIr}{\Delta x^4}\left[\frac{1}{16}\left(\theta_{i+1,j} - \theta_{i-1,j}\right)^4 - 3\left(\theta_{i+1,j} - 2\theta_{i,j} + \theta_{i-1,j}\right)^2\right.\right.$$

$$\left.- \left(\theta_{i+2,j} - 2\theta_{i+1,j} + 2\theta_{i-1,j} - \theta_{i-2,j}\right)\left(\theta_{i+1,j} - \theta_{i-1,j}\right)\right]$$

$$- \frac{EAr}{8\Delta x^3}\left[\left(U_{i+1,j} - U_{i-1,j}\right)\left(\theta_{i+1,j} - \theta_{i-1,j}\right)^2 + \frac{r^2}{4\Delta x}\left(\theta_{i+1,j} - \theta_{i-1,j}\right)^4\right]$$

$$+ \frac{I_p r\omega}{2\Delta x^2\Delta t}\left[\left(\theta_{i+1,j} - 2\theta_{i,j} + \theta_{i-1,j} - \theta_{i+1,j-2} + 2\theta_{i,j-2} - \theta_{i-1,j-2}\right)\right.$$

$$\left.- \frac{1}{4}\left(\theta_{i,j} - \theta_{i,j-2}\right)\left(\theta_{i+1,j} - \theta_{i-1,j}\right)^2\right]$$

$$\left.+m_p g\cos\theta_{i,j} + \frac{m_p r}{4\Delta t^2}\left(\theta_{i,j} - \theta_{i,j-2}\right)^2\right\} \tag{3.253}$$

$$\frac{I_p r\omega\left(\theta_{i+1,j} - \theta_{i-1,j}\right)}{4\Delta x^2\Delta t}\theta_{i+1,j+1}$$

$$+ \left[\frac{I_p r\omega\left(\theta_{i+1,j} - 2\theta_{i,j} + \theta_{i-1,j}\right)}{2\Delta x^2\Delta t} - \frac{m_p r}{\Delta t^2}\right]\theta_{i,j+1}$$

$$- \frac{I_p r\omega\left(\theta_{i+1,j} - \theta_{i-1,j}\right)}{4\Delta x^2\Delta t}\theta_{i-1,j+1}$$

$$= \frac{I_p r\omega\left(\theta_{i+1,j} - 2\theta_{i,j} + \theta_{i-1,j}\right)}{2\Delta x^2\Delta t}\theta_{i,j-1}$$

$$- \frac{I_p r\omega\left(\theta_{i+1,j} - \theta_{i-1,j}\right)}{4\Delta x^2\Delta t}\left(-\theta_{i+1,j-1} + \theta_{i-1,j-1}\right)$$

$$+ \frac{m_p r}{\Delta t^2}\left(-2\theta_{i,j} + \theta_{i,j-1}\right)$$

$$+ \frac{EIr}{\Delta x^4}\left[\left(\theta_{i+2,j} - 4\theta_{i+1,j} + 6\theta_{i,j} - 4\theta_{i-1,j} + \theta_{i-2,j}\right)\right.$$

$$\left.- \frac{3}{2}\left(\theta_{i+1,j} - \theta_{i-1,j}\right)^2\left(\theta_{i+1,j} - 2\theta_{i,j} + \theta_{i-1,j}\right)\right]$$

$$- \frac{EAr}{2\Delta x^3}\left[\left(U_{i+1,j} - U_{i-1,j}\right)\left(\theta_{i+1,j} - 2\theta_{i,j} + \theta_{i-1,j}\right)\right.$$

$$+ \left(U_{i+1,j} - 2U_{i,j} + U_{i-1,j}\right)\left(\theta_{i+1,j} - \theta_{i-1,j}\right)$$

$$+ \frac{3r^2}{4\Delta x}\left(\theta_{i+1,j} - 2\theta_{i,j} + \theta_{i-1,j}\right)\left(\theta_{i+1,j} - \theta_{i-1,j}\right)^2 \Big] + m_p g \sin\theta_{i,j}$$

$$+ \frac{\mathrm{sgn}\left(\theta_{i,j} - \theta_{i,j-1}\right) \mathrm{fr}\left|\theta_{i,j} - \theta_{i,j-2}\right|}{2\Delta t \sqrt{\left(\frac{U_{i,j} - U_{i,j-2}}{2\Delta t}\right)^2 + r^2 \left(\frac{\theta_{i,j} - \theta_{i,j-2}}{2\Delta t}\right)^2}}$$

$$\left\{ -\frac{\mathrm{EIr}}{\Delta x^4}\left[\frac{1}{16}\left(\theta_{i+1,j} - \theta_{i-1,j}\right)^4 - 3\left(\theta_{i+1,j} - 2\theta_{i,j} + \theta_{i-1,j}\right)^2 \right.\right.$$

$$\left. -\left(\theta_{i+2,j} - 2\theta_{i+1,j} + 2\theta_{i-1,j} - \theta_{i-2,j}\right)\left(\theta_{i+1,j} - \theta_{i-1,j}\right)\right]$$

$$- \frac{\mathrm{EAr}}{8\Delta x^3}\left[\left(U_{i+1,j} - U_{i-1,j}\right)\left(\theta_{i+1,j} - \theta_{i-1,j}\right)^2 + \frac{r^2}{4\Delta x}\left(\theta_{i+1,j} - \theta_{i-1,j}\right)^4 \right]$$

$$+ \frac{I_p r\omega}{2\Delta x^2 \Delta t}\left[\left(\theta_{i+1,j} - 2\theta_{i,j} + \theta_{i-1,j} - \theta_{i+1,j-2} + 2\theta_{i,j-2} - \theta_{i-1,j-2}\right)\right.$$

$$\left. - \frac{1}{4}\left(\theta_{i,j} - \theta_{i,j-2}\right)\left(\theta_{i+1,j} - \theta_{i-1,j}\right)^2 \right] + m_p g \cos\theta_{i,j}$$

$$+ \frac{m_p r}{4\Delta t^2}\left(\theta_{i,j} - \theta_{i,j-2}\right)^2 \right\} \tag{3.254}$$

$$F_{i,j} = -\frac{\mathrm{EA}}{2\Delta x}\left[\left(U_{i+1,j} - U_{i-1,j}\right) + \frac{r^2}{4\Delta x}\left(\theta_{i+1,j} - \theta_{i-1,j}\right)^2 \right] \tag{3.255}$$

$$N_{i,j} = -\frac{\mathrm{EIr}}{\Delta x^4}\left[\frac{1}{16}\left(\theta_{i+1,j} - \theta_{i-1,j}\right)^4 - 3\left(\theta_{i+1,j} - 2\theta_{i,j} + \theta_{i-1,j}\right)^2 \right.$$

$$\left. -\left(\theta_{i+2,j} - 2\theta_{i+1,j} + 2\theta_{i-1,j} - \theta_{i-2,j}\right)\left(\theta_{i+1,j} - \theta_{i-1,j}\right)\right]$$

$$- \frac{\mathrm{EAr}}{8\Delta x^3}\left[\left(U_{i+1,j} - U_{i-1,j}\right)\left(\theta_{i+1,j} - \theta_{i-1,j}\right)^2 + \frac{r^2}{4\Delta x}\left(\theta_{i+1,j} - \theta_{i-1,j}\right)^4 \right]$$

$$+ \frac{I_p r\omega}{2\Delta x^2 \Delta t}\left[\left(\theta_{i+1,j} - 2\theta_{i,j} + \theta_{i-1,j} - \theta_{i+1,j-2} + 2\theta_{i,j-2} - \theta_{i-1,j-2}\right)\right.$$

$$\left. - \frac{1}{4}\left(\theta_{i,j} - \theta_{i,j-2}\right)\left(\theta_{i+1,j} - \theta_{i-1,j}\right)^2 \right]$$

$$+ m_p g \cos\theta_{i,j} + \frac{m_p r}{4\Delta t^2}\left(\theta_{i,j} - \theta_{i,j-2}\right)^2 \tag{3.256}$$

The boundary conditions are the same given by Eqs. 3.159–3.164, while the initial conditions are the same given by Eqs. 3.171–3.174. The discretization can then be obtained by substituting the proper i and j in Eqs. 3.253–3.256.

3.2.4 Solution for Tripping Out

Differently from the model without friction, this time an analytical solution is not possible if friction is included. Repeating the three equations from the model:

$$EA\frac{\partial^2 u_x}{\partial x^2} - m_p\frac{\partial^2 u_x}{\partial t^2} - sgn\left(\frac{\partial u_x}{\partial t}\right)f\left[-m_p r\frac{\partial^3 u_x}{\partial x\partial t^2} + m_p g\right] = 0 \tag{3.257}$$

$$F_x(x,t) = -EA\frac{\partial u_x}{\partial x} \tag{3.258}$$

$$N(x,t) = -m_p r\frac{\partial^3 u_x}{\partial x\partial t^2} + m_p g \tag{3.259}$$

Discretizing Eqs. 3.257–3.259 and manipulating them:

$$U_{i,j+1} = 2U_{i,j} - U_{i,j-1} + \frac{EA\Delta t^2}{m_p \Delta x^2}\left(U_{i+1,j} - 2U_{i,j} + U_{i-1,j}\right)$$

$$- sgn\left(U_{i,j} - U_{i,j-1}\right)f\left[-\frac{r}{2\Delta x}\left(U_{i+1,j} - U_{i-1,j} - 2U_{i+1,j-1} + 2U_{i-1,j-1}\right.\right.$$

$$\left.\left. + U_{i+1,j-2} - U_{i-1,j-2}\right) + g\Delta t^2\right] \tag{3.260}$$

$$F_{i,j} = -\frac{EA}{2\Delta x}\left(U_{i+1,j} - U_{i-1,j}\right) \tag{3.261}$$

$$N_{i,j} = -\frac{m_p r}{2\Delta x\Delta t^2}\left(U_{i+1,j} - U_{i-1,j} - 2U_{i+1,j-1} + 2U_{i-1,j-1} + U_{i+1,j-2} - U_{i-1,j-2}\right)$$
$$+ m_p g$$

$$\tag{3.262}$$

The boundary conditions are given by Eqs. 3.185, 3.186, while the initial conditions are given by Eqs. 3.189, 3.190. Once again, the discretization can be obtained by substituting the proper i and j in Eqs. 3.260–3.262.

3.3 Model III: Columns Inside Slant Wells

The next step for modeling the column is to consider slant segments of well; up to this point, the model could only be applied to horizontal segments. In practice, even horizontal wells start its trajectory as vertical and have to gain angle before reaching the horizontal position. In addition, there are types of wells that do not even have horizontal segments, such as slant and S wells.

Figure 3.5 shows a scheme of a slant segment of well. In Fig. 3.5a, the segment is horizontal as in the previous section, while in Fig. 3.5b the segment has an inclination with angle α in respect to the vertical direction. Therefore, the model proposed on this section is a generalization of the previous one. The best way to characterize this problem is to keep using the coordinate system xyz, but now rotated to follow the well

Fig. 3.5   a Horizontal and **b** slant segment of well. The angle α defines the inclination in respect to the vertical direction

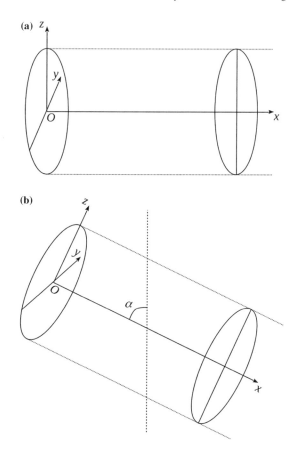

inclination. Observing Fig. 3.5b, it can be noted that the weight is different in this case; it now possesses a component on the x axis besides the plane yz; meanwhile, the normal contact force and the friction force still have the same components.

It is important to note that the previous hypotheses are still valid, which means that the column still remains always in contact with the wellbore, even if the well segment is not horizontal anymore. The implications of this hypothesis will be tested further on.

3.3.1   Model for Tripping in

The decomposition of the weight for the configuration given by Fig. 3.5b will be:

$$\vec{q}_p = m_p g dx \cos\alpha\,\hat{i} + m_p g dx \sin\alpha\cos\theta\,\hat{p} - m_p g dx \sin\alpha\sin\theta\,\hat{q} \qquad (3.263)$$

For $\alpha = 90°$ it can be noted that Eq. 3.263 reduces itself to Eq. 3.57 defined previously. Substituting Eqs. 3.58, 3.214 and 3.263 into Eq. 3.215:

$$\vec{f} = m_p g \cos\alpha\,\hat{i} + m_p g \sin\alpha \cos\theta\,\hat{p} - m_p g \sin\alpha \sin\theta\,\hat{q} - N\hat{p}$$
$$- \frac{fN}{\sqrt{\left(\frac{\partial u_x}{\partial t}\right)^2 + r^2\left(\frac{\partial\theta}{\partial t}\right)^2}}\left[\mathrm{sgn}\left(\frac{\partial u_x}{\partial t}\right)\left|\frac{\partial u_x}{\partial t}\right|\hat{i} + \mathrm{sgn}\left(\frac{\partial\theta}{\partial t}\right)r\left|\frac{\partial\theta}{\partial t}\right|\hat{q}\right] \tag{3.264}$$

$$\vec{f} = \left(m_p g \cos\alpha - \frac{\mathrm{sgn}\left(\frac{\partial u_x}{\partial t}\right)fN\left|\frac{\partial u_x}{\partial t}\right|}{\sqrt{\left(\frac{\partial u_x}{\partial t}\right)^2 + r^2\left(\frac{\partial\theta}{\partial t}\right)^2}}\right)\hat{i} + \left(m_p g \sin\alpha \cos\theta - N\right)\hat{p}$$
$$- \left(m_p g \sin\alpha \sin\theta + \frac{\mathrm{sgn}\left(\frac{\partial\theta}{\partial t}\right)fNr\left|\frac{\partial\theta}{\partial t}\right|}{\sqrt{\left(\frac{\partial u_x}{\partial t}\right)^2 + r^2\left(\frac{\partial\theta}{\partial t}\right)^2}}\right)\hat{q} \tag{3.265}$$

Substituting Eqs. 3.31, 3.49 and 3.265 into Eq. 3.74:

$$\frac{\partial F_x}{\partial x}\hat{i} + \left(\frac{\partial F_r}{\partial x} - F_\theta\frac{\partial\theta}{\partial x}\right)\hat{p} + \left(\frac{\partial F_\theta}{\partial x} + F_r\frac{\partial\theta}{\partial x}\right)\hat{q} - \left(m_p g \cos\alpha - \frac{\mathrm{sgn}\left(\frac{\partial u_x}{\partial t}\right)fN\left|\frac{\partial u_x}{\partial t}\right|}{\sqrt{\left(\frac{\partial u_x}{\partial t}\right)^2 + r^2\left(\frac{\partial\theta}{\partial t}\right)^2}}\right)\hat{i}$$
$$- \left(m_p g \sin\alpha \cos\theta - N\right)\hat{p} + \left(m_p g \sin\alpha \sin\theta + \frac{\mathrm{sgn}\left(\frac{\partial\theta}{\partial t}\right)fNr\left|\frac{\partial\theta}{\partial t}\right|}{\sqrt{\left(\frac{\partial u_x}{\partial t}\right)^2 + r^2\left(\frac{\partial\theta}{\partial t}\right)^2}}\right)\hat{q}$$
$$+ m_p\left[\frac{\partial^2 u_x}{\partial t^2}\hat{i} - r\left(\frac{\partial\theta}{\partial t}\right)^2\hat{p} + r\frac{\partial^2\theta}{\partial t^2}\hat{q}\right] = 0 \tag{3.266}$$

$$\left[\frac{\partial F_x}{\partial x} - m_p g \cos\alpha + \frac{\mathrm{sgn}\left(\frac{\partial u_x}{\partial t}\right)fN\left|\frac{\partial u_x}{\partial t}\right|}{\sqrt{\left(\frac{\partial u_x}{\partial t}\right)^2 + r^2\left(\frac{\partial\theta}{\partial t}\right)^2}} + m_p\frac{\partial^2 u_x}{\partial t^2}\right]\hat{i}$$
$$+ \left[\frac{\partial F_r}{\partial x} - F_\theta\frac{\partial\theta}{\partial x} + N - m_p g \sin\alpha \cos\theta - m_p r\left(\frac{\partial\theta}{\partial t}\right)^2\right]\hat{p}$$
$$+ \left[\frac{\partial F_\theta}{\partial x} + F_r\frac{\partial\theta}{\partial x} + m_p g \sin\alpha \sin\theta + \frac{\mathrm{sgn}\left(\frac{\partial\theta}{\partial t}\right)fNr\left|\frac{\partial\theta}{\partial t}\right|}{\sqrt{\left(\frac{\partial u_x}{\partial t}\right)^2 + r^2\left(\frac{\partial\theta}{\partial t}\right)^2}} + m_p r\frac{\partial^2\theta}{\partial t^2}\right]\hat{q} = 0 \tag{3.267}$$

Decomposing Eq. 3.267 into components:

$$\frac{\partial F_x}{\partial x} - m_p g \cos\alpha + \frac{\mathrm{sgn}\left(\frac{\partial u_x}{\partial t}\right)fN\left|\frac{\partial u_x}{\partial t}\right|}{\sqrt{\left(\frac{\partial u_x}{\partial t}\right)^2 + r^2\left(\frac{\partial\theta}{\partial t}\right)^2}} + m_p\frac{\partial^2 u_x}{\partial t^2} \tag{3.268}$$

$$\frac{\partial F_r}{\partial x} - F_\theta\frac{\partial\theta}{\partial x} + N - m_p g \sin\alpha \cos\theta - m_p r\left(\frac{\partial\theta}{\partial t}\right)^2 \tag{3.269}$$

$$\frac{\partial F_\theta}{\partial x} + F_r\frac{\partial \theta}{\partial x} + m_p g\,\sin\alpha\,\sin\theta + \frac{\mathrm{sgn}\!\left(\frac{\partial \theta}{\partial t}\right)fNr\left|\frac{\partial \theta}{\partial t}\right|}{\sqrt{\left(\frac{\partial u_x}{\partial t}\right)^2 + r^2\left(\frac{\partial \theta}{\partial t}\right)^2}} + m_p r\frac{\partial^2 \theta}{\partial t^2} \tag{3.270}$$

The expressions for F_x, F_r and F_θ from $\vec{F}$ do not modify from the previous cases. Therefore, substituting Eqs. 3.66, 3.92 and 3.95 into Eqs. 3.268–3.270, while using Eqs. 3.37, 3.38:

$$\frac{\partial}{\partial x}\left[-EA\frac{\partial u_x}{\partial x} - \frac{1}{2}EAr^2\left(\frac{\partial \theta}{\partial x}\right)^2\right] - m_p g\,\cos\alpha + \frac{\mathrm{sgn}\!\left(\frac{\partial u_x}{\partial t}\right)fN\left|\frac{\partial u_x}{\partial t}\right|}{\sqrt{\left(\frac{\partial u_x}{\partial t}\right)^2 + r^2\left(\frac{\partial \theta}{\partial t}\right)^2}} + m_p\frac{\partial^2 u_x}{\partial t^2} = 0 \tag{3.271}$$

$$\frac{\partial}{\partial x}\left[-3EIr\frac{\partial \theta}{\partial x}\frac{\partial^2 \theta}{\partial x^2} - I_p\omega r\frac{\partial^2 \theta}{\partial x\partial t}\right]$$

$$- \left[EIr\left[\frac{\partial^3 \theta}{\partial x^3} - \left(\frac{\partial \theta}{\partial x}\right)^3\right]\right]$$

$$+ \left[-EA\frac{\partial u_x}{\partial x} - \frac{1}{2}EAr^2\left(\frac{\partial \theta}{\partial x}\right)^2\right]r\frac{\partial \theta}{\partial x}$$

$$+ I_p\omega\left(-r\frac{\partial \theta}{\partial t}\frac{\partial \theta}{\partial x}\right)\right]\frac{\partial \theta}{\partial x} + N - m_p g\,\sin\alpha\,\cos\theta$$

$$- m_p r\left(\frac{\partial \theta}{\partial t}\right)^2 = 0 \tag{3.272}$$

$$\frac{\partial}{\partial x}\left[EIr\left[\frac{\partial^3 \theta}{\partial x^3} - \left(\frac{\partial \theta}{\partial x}\right)^3\right] + \left[-EA\frac{\partial u_x}{\partial x} - \frac{1}{2}EAr^2\left(\frac{\partial \theta}{\partial x}\right)^2\right]r\frac{\partial \theta}{\partial x}\right.$$

$$+ I_p\omega\left(-r\frac{\partial \theta}{\partial t}\frac{\partial \theta}{\partial x}\right)\right] + \left[-3EIr\frac{\partial \theta}{\partial x}\frac{\partial^2 \theta}{\partial x^2} - I_p\omega r\frac{\partial^2 \theta}{\partial x\partial t}\right]\frac{\partial \theta}{\partial x}$$

$$+ m_p g\,\sin\alpha\,\sin\theta + \frac{\mathrm{sgn}\!\left(\frac{\partial \theta}{\partial t}\right)fNr\left|\frac{\partial \theta}{\partial t}\right|}{\sqrt{\left(\frac{\partial u_x}{\partial t}\right)^2 + r^2\left(\frac{\partial \theta}{\partial t}\right)^2}} + m_p r\frac{\partial^2 \theta}{\partial t^2}$$

$$= 0 \tag{3.273}$$

Manipulating Eqs. 3.271–3.273:

$$EA\frac{\partial^2 u_x}{\partial x^2} - m_p\frac{\partial^2 u_x}{\partial t^2} + EAr^2\frac{\partial \theta}{\partial x}\frac{\partial^2 \theta}{\partial x^2} - \frac{\mathrm{sgn}\!\left(\frac{\partial u_x}{\partial t}\right)fN\left|\frac{\partial u_x}{\partial t}\right|}{\sqrt{\left(\frac{\partial u_x}{\partial t}\right)^2 + r^2\left(\frac{\partial \theta}{\partial t}\right)^2}} + m_p g\,\cos\alpha = 0 \tag{3.274}$$

$$N = -EIr\left[\left(\frac{\partial\theta}{\partial x}\right)^4 - 3\left(\frac{\partial^2\theta}{\partial x^2}\right)^2 - 4\frac{\partial^3\theta}{\partial x^3}\frac{\partial\theta}{\partial x}\right] - EAr\left[\frac{\partial u_x}{\partial x}\left(\frac{\partial\theta}{\partial x}\right)^2 + \frac{1}{2}r^2\left(\frac{\partial\theta}{\partial x}\right)^4\right] +$$

$$I_p r\omega\left[\frac{\partial^3\theta}{\partial x^2\partial t} - \frac{\partial\theta}{\partial t}\left(\frac{\partial\theta}{\partial x}\right)^2\right] + m_p g \sin\alpha \cos\theta + m_p r\left(\frac{\partial\theta}{\partial t}\right)^2 \qquad (3.275)$$

$$EIr\left[\frac{\partial^4\theta}{\partial x^4} - 6\left(\frac{\partial\theta}{\partial x}\right)^2\frac{\partial^2\theta}{\partial x^2}\right] - EAr\left[\frac{\partial u_x}{\partial x}\frac{\partial^2\theta}{\partial x^2} + \frac{\partial^2 u_x}{\partial x^2}\frac{\partial\theta}{\partial x} + \frac{3}{2}r^2\frac{\partial^2\theta}{\partial x^2}\left(\frac{\partial\theta}{\partial x}\right)^2\right]$$

$$- I_p r\omega\left[2\frac{\partial^2\theta}{\partial x\partial t}\frac{\partial\theta}{\partial x} + \frac{\partial\theta}{\partial t}\frac{\partial^2\theta}{\partial x^2}\right] + m_p g \sin\alpha \sin\theta + m_p r\frac{\partial^2\theta}{\partial t^2} + \frac{\text{sgn}\left(\frac{\partial\theta}{\partial t}\right)fNr\left|\frac{\partial\theta}{\partial t}\right|}{\sqrt{\left(\frac{\partial u_x}{\partial t}\right)^2 + r^2\left(\frac{\partial\theta}{\partial t}\right)^2}} = 0$$
$$(3.276)$$

Lastly, substituting the normal contact force given by Eq. 3.275 into Eqs. 3.274 and 3.276:

$$EA\frac{\partial^2 u_x}{\partial x^2} - m_p\frac{\partial^2 u_x}{\partial t^2} + EAr^2\frac{\partial\theta}{\partial x}\frac{\partial^2\theta}{\partial x^2} \qquad (3.277)$$

$$- \frac{\text{sgn}\left(\frac{\partial u_x}{\partial t}\right)f\left|\frac{\partial u_x}{\partial t}\right|}{\sqrt{\left(\frac{\partial u_x}{\partial t}\right)^2 + r^2\left(\frac{\partial\theta}{\partial t}\right)^2}}\left\{-EIr\left[\left(\frac{\partial\theta}{\partial x}\right)^4 - 3\left(\frac{\partial^2\theta}{\partial x^2}\right)^2 - 4\frac{\partial^3\theta}{\partial x^3}\frac{\partial\theta}{\partial x}\right]\right.$$

$$- EAr\left[\frac{\partial u_x}{\partial x}\left(\frac{\partial\theta}{\partial x}\right)^2 + \frac{1}{2}r^2\left(\frac{\partial\theta}{\partial x}\right)^4\right] + I_p r\omega\left[\frac{\partial^3\theta}{\partial x^2\partial t} - \frac{\partial\theta}{\partial t}\left(\frac{\partial\theta}{\partial x}\right)^2\right]$$

$$\left. + m_p g \sin\alpha \cos\theta + m_p r\left(\frac{\partial\theta}{\partial t}\right)^2\right\} + m_p g \cos\alpha = 0$$

$$EIr\left[\frac{\partial^4\theta}{\partial x^4} - 6\left(\frac{\partial\theta}{\partial x}\right)^2\frac{\partial^2\theta}{\partial x^2}\right] \qquad (3.278)$$

$$- EAr\left[\frac{\partial u_x}{\partial x}\frac{\partial^2\theta}{\partial x^2} + \frac{\partial^2 u_x}{\partial x^2}\frac{\partial\theta}{\partial x} + \frac{3}{2}r^2\frac{\partial^2\theta}{\partial x^2}\left(\frac{\partial\theta}{\partial x}\right)^2\right]$$

$$- I_p r\omega\left[2\frac{\partial^2\theta}{\partial x\partial t}\frac{\partial\theta}{\partial x} + \frac{\partial\theta}{\partial t}\frac{\partial^2\theta}{\partial x^2}\right] + m_p g \sin\alpha \sin\theta + m_p r\frac{\partial^2\theta}{\partial t^2}$$

$$+ \frac{\text{sgn}\left(\frac{\partial\theta}{\partial t}\right)fr\left|\frac{\partial\theta}{\partial t}\right|}{\sqrt{\left(\frac{\partial u_x}{\partial t}\right)^2 + r^2\left(\frac{\partial\theta}{\partial t}\right)^2}}\left\{-EIr\left[\left(\frac{\partial\theta}{\partial x}\right)^4 - 3\left(\frac{\partial^2\theta}{\partial x^2}\right)^2 - 4\frac{\partial^3\theta}{\partial x^3}\frac{\partial\theta}{\partial x}\right]\right.$$

$$- EAr\left[\frac{\partial u_x}{\partial x}\left(\frac{\partial\theta}{\partial x}\right)^2 + \frac{1}{2}r^2\left(\frac{\partial\theta}{\partial x}\right)^4\right] + I_p r\omega\left[\frac{\partial^3\theta}{\partial x^2\partial t} - \frac{\partial\theta}{\partial t}\left(\frac{\partial\theta}{\partial x}\right)^2\right]$$

$$\left. + m_p g \sin\alpha \cos\theta + m_p r\left(\frac{\partial\theta}{\partial t}\right)^2\right\} = 0$$

The problem consists in solving Eqs. 3.275, 3.277 and 3.278 in order to find the axial and angular displacements and the normal contact force, besides Eq. 3.66 for the axial force.

3.3.2 Model for Tripping Out

For the case of tripping out, the weight decomposition according to Fig. 3.5b will be given by:

$$\vec{q}_p = m_p g dx \cos \alpha \, \hat{i} - m_p g dx \sin \alpha \, \hat{k} \tag{3.279}$$

For $\alpha = 90°$, Eq. 3.279 reduces itself to Eq. 3.125 previously defined. Substituting Eqs. 3.126, 3.235 and 3.279 into Eq. 3.236:

$$\vec{f} = m_p g \cos \alpha \, \hat{i} - m_p g \sin \alpha \, \hat{k} + N \hat{k} - \text{sgn} \left(\frac{\partial u_x}{\partial t} \right) f N \hat{i} \tag{3.280}$$

$$\vec{f} = \left(m_p g \cos \alpha - \text{sgn} \left(\frac{\partial u_x}{\partial t} \right) f N \right) \hat{i} + \left(-m_p g \sin \alpha + N \right) \hat{k} \tag{3.281}$$

Substituting Eqs. 3.112, 3.122 and 3.281 into Eq. 3.134:

$$\frac{\partial F_x}{\partial x} \hat{i} + \frac{\partial F_y}{\partial x} \hat{j} + \frac{\partial F_z}{\partial x} \hat{k} + \text{sgn} \left(\frac{\partial u_x}{\partial t} \right) f N \hat{i}$$
$$- \left[\left(m_p g \cos \alpha - \text{sgn} \left(\frac{\partial u_x}{\partial t} \right) f N \right) \hat{i} + \left(-m_p g \sin \alpha + N \right) \hat{k} \right] + m_p \left[\frac{\partial^2 u_x}{\partial t^2} \hat{i} \right] = 0 \tag{3.282}$$

$$\left(\frac{\partial F_x}{\partial x} + m_p \frac{\partial^2 u_x}{\partial t^2} - m_p g \cos \alpha + \text{sgn} \left(\frac{\partial u_x}{\partial t} \right) f N \right) \hat{i} + \left(\frac{\partial F_y}{\partial x} \right) \hat{j}$$
$$+ \left(\frac{\partial F_z}{\partial x} + m_p g \sin \alpha - N \right) \hat{k} = 0 \tag{3.283}$$

Separating Eq. 3.283 into its components:

$$\frac{\partial F_x}{\partial x} + m_p \frac{\partial^2 u_x}{\partial t^2} - m_p g \cos \alpha + \text{sgn} \left(\frac{\partial u_x}{\partial t} \right) f N = 0 \tag{3.284}$$

$$\frac{\partial F_y}{\partial x} = 0 \tag{3.285}$$

$$\frac{\partial F_z}{\partial x} + m_p g \sin \alpha - N = 0 \tag{3.286}$$

Once again, the expressions for F_x, F_y and F_z from $\overrightarrow{F}$ do not modify from the previous cases. Therefore, substituting Eqs. 3.131, 3.147 and 3.148 into Eqs. 3.284–3.286, two equations of motion are obtained since Eq. 3.285 is irrelevant:

$$\frac{\partial}{\partial x}\left[-EA\frac{\partial u_x}{\partial x}\right] + m_p\frac{\partial^2 u_x}{\partial t^2} - m_p g \cos\alpha + \text{sgn}\left(\frac{\partial u_x}{\partial t}\right) fN = 0 \qquad (3.287)$$

$$\frac{\partial}{\partial x}\left[-m_p r\frac{\partial^2 u_x}{\partial t^2}\right] + m_p g \sin\alpha - N = 0 \qquad (3.288)$$

Manipulating Eqs. 3.287, 3.288:

$$EA\frac{\partial^2 u_x}{\partial x^2} - m_p\frac{\partial^2 u_x}{\partial t^2} + m_p g \cos\alpha - \text{sgn}\left(\frac{\partial u_x}{\partial t}\right) fN = 0 \qquad (3.289)$$

$$N = -m_p r\frac{\partial^3 u_x}{\partial x \partial t^2} + m_p g \sin\alpha \qquad (3.290)$$

Finally, substituting Eq. 3.290 into Eq. 3.289:

$$EA\frac{\partial^2 u_x}{\partial x^2} - m_p\frac{\partial^2 u_x}{\partial t^2} + m_p g \cos\alpha - \text{sgn}\left(\frac{\partial u_x}{\partial t}\right) f\left[-m_p r\frac{\partial^3 u_x}{\partial x \partial t^2} + m_p g \sin\alpha\right] = 0$$
$$(3.291)$$

The problem consists of solving Eqs. 3.290, 3.291 to find the axial displacement and normal contact force, besides Eq. 3.131 for the axial force.

3.3.3 Solution for Tripping in

Discretizing Eqs. 3.66, 3.275, 3.277 and 3.278 and manipulating them:

$$U_{i,j+1} = 2U_{i,j} - U_{i,j-1} + \frac{EA\Delta t^2}{m_p \Delta x^2}\left(U_{i+1,j} - 2U_{i,j} + U_{i-1,j}\right)$$

$$+ \frac{EAr^2\Delta t^2}{2m_p\Delta x^3}\left(\theta_{zi+1,j} - \theta_{i-1,j}\right)\left(\theta_{i+1,j} - 2\theta_{i,j} + \theta_{i-1,j}\right)$$

$$- \frac{\mathrm{sgn}(U_{i,j} - U_{i,j-1})f\Delta t\left|U_{i,j} - U_{i,j-2}\right|}{2m_p\sqrt{\left(\frac{U_{i,j}-U_{i,j-2}}{2\Delta t}\right)^2 + r^2\left(\frac{\theta_{i,j}-\theta_{i,j-2}}{2\Delta t}\right)}}$$

$$\left\{-\frac{EIr}{\Delta x^4}\left[\frac{1}{16}\left(\theta_{i+1,j} - \theta_{i-1,j}\right)^4 - 3\left(\theta_{i+1,j} - 2\theta_{i,j} + \theta_{i-1,j}\right)^2\right.\right.$$

$$\left.-\left(\theta_{i+2,j} - 2\theta_{i+1,j} + 2\theta_{i-1,j} - \theta_{i-2,j}\right)\left(\theta_{i+1,j} - \theta_{i-1,j}\right)\right]$$

$$- \frac{EAr}{8\Delta x^3}\left[\left(U_{i+1,j} - U_{i-1,j}\right)\left(\theta_{i+1,j} - \theta_{i-1,j}\right)^2 + \frac{r^2}{4\Delta x}\left(\theta_{i+1,j} - \theta_{i-1,j}\right)^4\right]$$

$$+ \frac{I_p r\omega}{2\Delta x^2\Delta t}\left[\left(\theta_{i+1,j} - 2\theta_{i,j} + \theta_{i-1,j} - \theta_{i+1,j-2} + 2\theta_{i,j-2} - \theta_{i-1,j-2}\right)\right.$$

$$\left.-\frac{1}{4}\left(\theta_{i,j} - \theta_{i,j-2}\right)\left(\theta_{i+1,j} - \theta_{i-1,j}\right)^2\right]$$

$$\left.+ m_p g\sin\alpha\cos\theta_{i,j} + \frac{m_p r}{4\Delta t^2}\left(\theta_{i,j} - \theta_{i,j-2}\right)^2\right\} + \Delta t^2 g\cos\alpha \qquad (3.292)$$

$$\frac{I_p r\omega\left(\theta_{i+1,j} - \theta_{i-1,j}\right)}{4\Delta x^2\Delta t}\theta_{i+1,j+1} + \left[\frac{I_p r\omega\left(\theta_{i+1,j} - 2\theta_{i,j} + \theta_{i-1,j}\right)}{2\Delta x^2\Delta t} - \frac{m_p r}{\Delta t^2}\right]\theta_{i,j+1}$$

$$- \frac{I_p r\omega\left(\theta_{i+1,j} - \theta_{i-1,j}\right)}{4\Delta x^2\Delta t}\theta_{i-1,j+1} = \frac{I_p r\omega\left(\theta_{i+1,j} - 2\theta_{i,j} + \theta_{i-1,j}\right)}{2\Delta x^2\Delta t}\theta_{i,j-1}$$

$$- \frac{I_p r\omega\left(\theta_{i+1,j} - \theta_{i-1,j}\right)}{4\Delta x^2\Delta t}\left(-\theta_{i+1,j-1} + \theta_{i-1,j-1}\right) + \frac{m_p r}{\Delta t^2}\left(-2\theta_{i,j} + \theta_{i,j-1}\right)$$

$$+ \frac{EIr}{\Delta x^4}\left[\left(\theta_{i+2,j} - 4\theta_{i+1,j} + 6\theta_{i,j} - 4\theta_{i-1,j} + \theta_{i-2,j}\right)\right.$$

$$\left.-\frac{3}{2}\left(\theta_{i+1,j} - \theta_{i-1,j}\right)^2\left(\theta_{i+1,j} - 2\theta_{i,j} + \theta_{i-1,j}\right)\right]$$

$$- \frac{EAr}{2\Delta x^3}\left[\left(U_{i+1,j} - U_{i-1,j}\right)\left(\theta_{i+1,j} - 2\theta_{i,j} + \theta_{i-1,j}\right)\right.$$

$$+\left(U_{i+1,j} - 2U_{i,j} + U_{i-1,j}\right)\left(\theta_{i+1,j} - \theta_{i-1,j}\right)$$

$$\left.+\frac{3r^2}{4\Delta x}\left(\theta_{i+1,j} - 2\theta_{i,j} + \theta_{i-1,j}\right)\left(\theta_{i+1,j} - \theta_{i-1,j}\right)^2\right]$$

$$+ m_p g\sin\alpha\sin\theta_{i,j}$$

$$+ \frac{\mathrm{sgn}\left(\theta_{i,j} - \theta_{i,j-1}\right)\mathrm{fr}\left|\theta_{i,j} - \theta_{i,j-2}\right|}{2\Delta t \sqrt{\left(\frac{U_{i,j} - U_{i,j-2}}{2\Delta t}\right)^2 + r^2\left(\frac{\theta_{i,j} - \theta_{i,j-2}}{2\Delta t}\right)^2}}$$

$$\begin{aligned}
\Big\{ &- \frac{EIr}{\Delta x^4}\left[\frac{1}{16}\left(\theta_{i+1,j} - \theta_{i-1,j}\right)^4 - 3\left(\theta_{i+1,j} - 2\theta_{i,j} + \theta_{i-1,j}\right)^2\right. \\
&\left. - \left(\theta_{i+2,j} - 2\theta_{i+1,j} + 2\theta_{i-1,j} - \theta_{i-2,j}\right)\left(\theta_{i+1,j} - \theta_{i-1,j}\right)\right] \\
&- \frac{EAr}{8\Delta x^3}\left[\left(U_{i+1,j} - U_{i-1,j}\right)\left(\theta_{i+1,j} - \theta_{i-1,j}\right)^2 + \frac{r^2}{4\Delta x}\left(\theta_{i+1,j} - \theta_{i-1,j}\right)^4\right] \\
&+ \frac{I_p r\omega}{2\Delta x^2 \Delta t}\left[\left(\theta_{i+1,j} - 2\theta_{i,j} + \theta_{i-1,j} - \theta_{i+1,j-2} + 2\theta_{i,j-2} - \theta_{i-1,j-2}\right)\right. \\
&\left. - \frac{1}{4}\left(\theta_{i,j} - \theta_{i,j-2}\right)\left(\theta_{i+1,j} - \theta_{i-1,j}\right)^2\right] \\
&+ m_p g \sin \alpha \cos \theta_{i,j} + \frac{m_p r}{4\Delta t^2}\left(\theta_{i,j} - \theta_{i,j-2}\right)^2 \Big\}
\end{aligned} \tag{3.293}$$

$$F_{i,j} = -\frac{EA}{2\Delta x}\left[\left(U_{i+1,j} - U_{i-1,j}\right) + \frac{r^2}{4\Delta x}\left(\theta_{i+1,j} - \theta_{i-1,j}\right)^2\right] \tag{3.294}$$

$$\begin{aligned}
N_{i,j} = &- \frac{EIr}{\Delta x^4}\left[\frac{1}{16}\left(\theta_{i+1,j} - \theta_{i-1,j}\right)^4 - 3\left(\theta_{i+1,j} - 2\theta_{i,j} + \theta_{i-1,j}\right)^2\right. \\
&\left. - \left(\theta_{i+2,j} - 2\theta_{i+1,j} + 2\theta_{i-1,j} - \theta_{i-2,j}\right)\left(\theta_{i+1,j} - \theta_{i-1,j}\right)\right] \\
&- \frac{EAr}{8\Delta x^3}\left[\left(U_{i+1,j} - U_{i-1,j}\right)\left(\theta_{i+1,j} - \theta_{i-1,j}\right)^2 + \frac{r^2}{4\Delta x}\left(\theta_{i+1,j} - \theta_{i-1,j}\right)^4\right] \\
&+ \frac{I_p r\omega}{2\Delta x^2 \Delta t}\left[\left(\theta_{i+1,j} - 2\theta_{i,j} + \theta_{i-1,j} - \theta_{i+1,j-2} + 2\theta_{i,j-2} - \theta_{i-1,j-2}\right)\right. \\
&\left. - \frac{1}{4}\left(\theta_{i,j} - \theta_{i,j-2}\right)\left(\theta_{i+1,j} - \theta_{i-1,j}\right)^2\right] \\
&+ m_p g \sin \alpha \cos \theta_{i,j} + \frac{m_p r}{4\Delta t^2}\left(\theta_{i,j} - \theta_{i,j-2}\right)^2
\end{aligned} \tag{3.295}$$

The boundary conditions and initial conditions are the same ones from the previous cases. The discretization is obtained using Eqs. 3.292–3.295.

3.3.4   Solution for Tripping Out

Discretizing Eqs. 3.131, 3.290 and 3.291 and manipulating them:

$$U_{i,j+1} = 2U_{i,j} - U_{i,j-1} + \frac{EA\Delta t^2}{m_p \Delta x^2}\left(U_{i+1,j} - 2U_{i,j} + U_{i-1,j}\right) + g\Delta t^2 \cos\alpha$$

$$- \operatorname{sgn}\left(U_{i,j} - U_{i,j-1}\right)f\left[-\frac{r}{2\Delta x}\left(U_{i+1,j} - U_{i-1,j} - 2U_{i+1,j-1} + 2U_{i-1,j-1}\right.\right.$$

$$\left.\left. + U_{i+1,j-2} - U_{i-1,j-2}\right) + g\Delta t^2 \sin\alpha\right]$$

$$\hspace{8cm}(3.296)$$

$$F_{i,j} = -\frac{EA}{2\Delta x}\left(U_{i+1,j} - U_{i-1,j}\right) \hspace{3cm}(3.297)$$

$$N_{i,j} = -\frac{m_p r}{2\Delta x \Delta t^2}\left(U_{i+1,j} - U_{i-1,j} - 2U_{i+1,j-1} + 2U_{i-1,j-1} + U_{i+1,j-2} - U_{i-1,j-2}\right)$$

$$+ m_p g \sin\alpha$$

$$\hspace{8cm}(3.298)$$

The boundary conditions and initial conditions are the same ones from the previous cases. The discretization is obtained using Eqs. 3.296–3.298.

3.4   Model IV: Columns Inside Offshore Wells

Up until now, all models considered the same boundary conditions: fixed at $x = 0$ and free at $x = L$. However, such boundary conditions are not enough to describe the behavior of columns inside offshore wells. On offshore environments, the environment causes displacements on the platform or vessel, in which the columns are attached; consequently, the column will displace as well due to this motion. These displacements are the consequence of environmental loads, which can be waves, currents and/or wind. The models developed so far can still be used for columns inside onshore wells or attached to fixed platforms, since on these cases, the environment cannot cause any meaningful loads and thus the column can be considered fixed. Figure 3.6 shows the types of displacements—also known as degrees of freedom—that can occur on a floating vessel.

Observing Fig. 3.6, it can be concluded that the vessel will be subjected to three linear displacements on the x, y and z axis—surge, heave and sway, respectively—and three angular displacements around the x, y and z axis—roll, yaw and pitch, respectively. These six motions will be transmitted to any column that is coupled to the vessel. Due to the hypotheses adopted on this book, only the linear displacements can be considered. Regarding them, the most relevant effect is the heave motion; the amplitudes of surge and sway are too small in comparison, since the vessel can absorb these motions due to its dynamic positioning system (DPS). Therefore, only the vessel heave is considered and it is transmitted to the column as a boundary condition for the axial displacement u_x. Chung and Whitney (1981) already used this

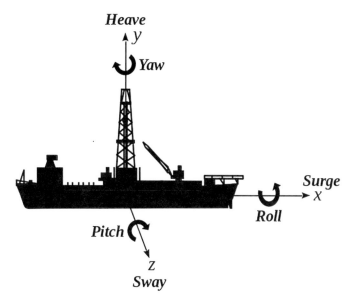

Fig. 3.6 Degrees of freedom of a floating vessel

methodology. Lastly, it is important to note that the vessel displacement amplitudes are not transmitted entirely to the coupled column. The transmitted amount depends on the motion frequency; the response amplitude operators (RAOs) of the vessel give this information. These operators exist for each one of the six degrees of freedom and are characteristic of each vessel. To simplify the analysis, it is considered that the displacement used as a boundary condition is already the column displacement, thus eliminating the need and usage of a RAO.

The boundary condition to be defined here is only for the axial displacement at $x = 0$; this means that from the six boundary conditions defined previously—three for each end, with one being for u_x, one for θ and one for $\partial^2\theta/\partial x^2$—only one will be changed: u_x at $x = 0$. For u_x at $x = 0$:

$$u(0, t) = U_h \sin \omega_h t \tag{3.299}$$

where U_h is the heave amplitude and ω_h is the heave angular frequency. Discretizing Eq. 3.299:

$$U_{0,j} = U_h \sin\left[(j - 2)\omega_h \Delta t\right] \tag{3.300}$$

The remaining equations for this model are the same developed previously, but remembering that the values for $U_{0,j}$ are no longer zero, thus they do not disappear from the $i = 1$ and $i = 2$ equations. The solutions are valid only starting at $j = 3$, since for $j = 1$ and $j = 2$ the initial conditions are applied instead; that is why on Eq. 3.300 there is $j-2$ instead of only j.

3.5   Summary

This chapter:

- Introduces models to study dynamic column buckling:
    - Model I, for frictionless columns inside purely horizontal segments of wells;
    - Model II, for columns with friction inside purely horizontal segments of wells;
    - Model III, for columns with friction inside directional wells, with trajectories fully described, on onshore environments;
    - Model IV, for columns with friction inside directional wells, with trajectories fully described, on offshore environments.

References

Chung JS, Whitney AK (1981) Dynamic vertical stretching oscillation of an 18000 ft ocean mining pipe. In: Offshore technology conference, 4–7 May, Houston, Texas. OTC 4092-MS. http://dx. doi.org/10.4043/4092-MS

Gao G, Miska S (2009) Effects of boundary conditions and friction on static buckling of pipe in a horizontal well. SPE J 14(4): 782–796, SPE 111511-PA. http://dx.doi.org/10.2118/111511-PA

Gao G, Miska S (2010) Dynamic buckling and snaking motion of rotating drilling pipe in a horizontal well. SPE J 15(3): 867–877, SPE 113883-PA. http://dx.doi.org/10.2118/113883-PA

Mitchell RF (1986) Simple frictional analysis of helical buckling of tubing. SPE Drill Eng 1(6): 457–465, SPE 13064-PA. http://dx.doi.org/10.2118/13064-PA

Mitchell RF (1996) Comprehensive analysis of buckling with friction. SPE Drill Completion 11(3): 178–184, SPE 29457-PA. http://dx.doi.org/10.2118/29457-PA

Mitchell RF (2007) The effect of friction on initial buckling of tubing and flowlines. SPE Drill Completions 22(2): 112–118, SPE 99099-PA. http://dx.doi.org/10.2118/99099-PA

Chapter 4
Case Study

A methodology to solve the equations and to analyze the results obtained is presented here. A summary of the equations and variables involved is made and a case study—based on real data—is also presented.

4.1 Methodology

As shown in Chap. 3, four models were developed to study the column dynamic buckling problem. Figure 4.1 summarizes the variables of the problem, while Table 4.1 summarizes the equations needed for each model, considering that there is a set of equations for tripping in and for tripping out inside each model.

After doing the discretization using an implicit method, the axial displacement can be calculated independently for each point, whereas the angular displacement results into a system of algebraic equations. After both displacements are calculated, the axial, normal contact, and friction forces can then be obtained. Therefore, the problems range from three to five unknowns. Table 4.2 summarizes all the variables present in each model. For tripping out cases, the angular displacement θ does not exist, while for Model I the friction force components $\vec{F}_{f1}$ and $\vec{F}_{f2}$ do not exist. As seen in Table 4.2, there are 34 variables to be analyzed.

In order to simulate each model, a data set is also needed. While some properties remain the same across multiple scenarios—such as the material properties—other inputs may change depending on the operation—such as the diameters. In this book, a single scenario was considered: a tubing string inside a cased hole. Table 4.3 presents all the data used. It is important to point that the casing is considered to reach from the bottom throughout the whole well until the top; this implicates that a liner is not used. In addition, as mentioned before, all the models were developed for a drill string, which rotates; however, the scenario here is for a completion operation, in which the column does not rotate. Also, some data is exclusive to certain models:

© The Author(s) 2018
M. A. Jaculli and J. R. P. Mendes, *Dynamic Buckling of Columns Inside Oil Wells*,
SpringerBriefs in Petroleum Geoscience & Engineering,
https://doi.org/10.1007/978-3-319-91208-0_4

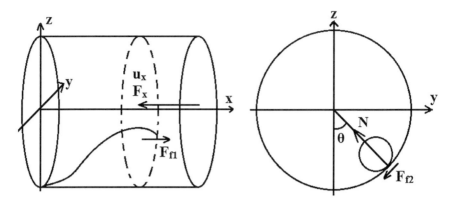

Fig. 4.1 Well schematic with all the variables from the problem

Table 4.1 Summary of all equations needed to solve each model

Model I	Tripping in	Eqs. 3.66, 3.81, 3.101 and 3.102
	Tripping out	Eqs. 3.131, 3.141 and 3.150
Model II	Tripping in	Eqs. 3.66, 3.101, 3.214, 3.230 and 3.231
	Tripping out	Eqs. 3.131, 3.150, 3.235 and 3.248
Model III	Tripping in	Eqs. 3.66, 3.214, 3.275, 3.277 and 3.278
	Tripping out	Eqs. 3.131, 3.235, 3.290 and 3.291
Model IV	Tripping in	Eqs. 3.66, 3.214, 3.275, 3.277 and 3.278
	Tripping out	Eqs. 3.131, 3.235, 3.290 and 3.291

Table 4.2 Summary of all variables of the problem. The X marks if a certain variable appears on the corresponding model

		u_x	θ	F_x	N	F_f
Model I	Tripping in	X	X	X	X	
	Tripping out	X		X	X	
Model II	Tripping in	X	X	X	X	X
	Tripping out	X		X	X	X
Model III	Tripping in	X	X	X	X	X
	Tripping out	X		X	X	X
Model IV	Tripping in	X	X	X	X	X
	Tripping out	X		X	X	X

the dynamic friction force appears only from Model II onwards since Model I does not have the friction force; the heave is modeled with constant amplitude, constant angular frequency, and in phase with the rest of the system, appearing only in Model IV.

Table 4.3 Data used to simulate the tubing-casing scenario

Property	Value
Tubing string inner diameter	5.791 in (0.1471 m)
Tubing string outer diameter	6.625 in (0.1683 m)
Inner diameter of a 10 ¾" casing	9.56 in (0.2428 m)
Young's modulus	210 GPa
Material density	7850 kg/m^3
Gravitational acceleration	9.81 m/s^2
Rotation angular frequency	0 rad/s (not rotating)
Dynamic dry friction coefficient (for Models II, III and IV)	0.1
Heave amplitude (for Model IV only)	0.5 m
Heave angular frequency (for Model IV only)	1 rad/s
Space discretization interval	10 m
Time discretization interval	0.0002 s
Space domain	1000 m (Models I and II) Fig. 4.2 (Models III and IV)
Time domain	10 s (20 s for Model IV)

Last, for Models III and IV, which consider the well inclination, it is possible to define a well trajectory. Although both models only apply for a segment with constant inclination, the key idea here is that a small curved segment of wellbore can be approximated as a segment with constant inclination and thus all equations can be applied. By joining together several of these small segments—with each one of them having their own angle of inclination α according to their position on the well—any given well trajectory can be discretized. Since the solution will be obtained using the finite differences method, a different inclination angle can be assigned to each measured depth and the simulation can calculate the displacements and forces for the whole column inside the well. In this book, a single well trajectory was considered: a horizontal well with two build-ups, as seen in Fig. 4.2.

The well starts vertical and must reach the objective, which is at a true vertical depth of 2068 m and has a horizontal departure of 1640 m. The KOP (kickoff point) must be located at a true vertical depth of 945 m. The well starts building its angle with a rate of 2°/30 m until it reaches 55°. After the slant segment reaches a true vertical depth of 1968 m, the well starts building its angle once more, now with a rate of 3°/30 m until it reaches 90°. After the horizontal position is reached, the well continues with a purely horizontal segment of 500 m. The remaining values, which are calculated, are also shown in Fig. 4.2. The first curved segment has a radius of 860 m, while the second one has a radius of 573 m. The horizontal departure just

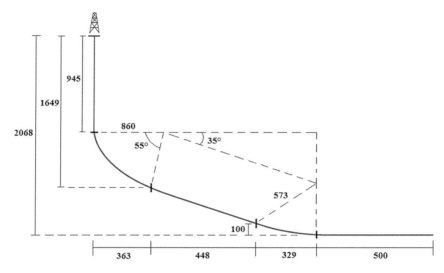

Fig. 4.2 Well trajectory for a horizontal well with two build-ups. All lengths are in meters

before the horizontal segment starts is 1140 m. The total measured depth of the well is 3170 m.

Since for Models I and II the well inclination is irrelevant—because the model is only valid for horizontal segments—a length of L= 1000 m will be used instead to obtain preliminary results. The trajectories will then be applied to Models III and IV. For comparison purposes between models, the same column length will be used for Models III and IV, when in reality Model IV should have an extra vertical segment at the beginning, representing the column length connecting the vessel with the wellhead along the water depth. The model can still be applied in this case since the column is still constrained—now by a marine riser—through the whole water depth.

4.2 Results and Discussion

Using the variables shown in Table 4.2, several comparisons can be drawn: besides comparing tripping in and tripping out for each model and testing several inclinations for Models III and IV, the models can be compared between themselves. For example, Models I and II can be compared to see the effect of friction, while Models III and IV can be compared to see the effect of the heave. Finally, each graph can be made as a function of either time or position; the former would give the behavior of a variable on a fixed point of the column while the latter would give the behavior of a variable for all points on a fixed time instant.

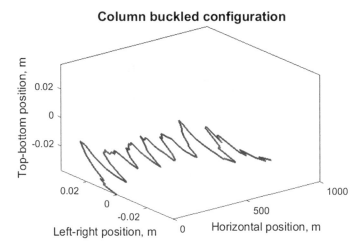

Fig. 4.3 Column buckled configuration, given by the positions, in meters, in the three axes. The displacements used are for the time instant $t = 5$ s

Figure 4.3 shows the column buckled configuration for tripping in using Model I. The column has been reduced to a single line, in which each point is the center of the cross section. Equation 3.1 gives the position of each cross section center. Data was taken for $t = 5$ s. For this graph only, the initial condition for θ was set at 0.7 rad to allow a better visualization.

It can be seen in Fig. 4.3 that the column remains on the lowest portion of the well—represented by the negative values on the top-bottom position axis—but can reach values around the halfway mark of this portion, which would be $r_c/2$.

Figure 4.4 shows a 3D graph for the horizontal displacement as a function of both horizontal position and time for Model II. By looking at the time axis, it can be seen that the displacement dissipates over time due to friction; meanwhile, by looking at the position axis, the displacement distribution along the length can be seen for a fixed time instant.

Figure 4.5 shows a comparison of the horizontal displacement between tripping in and tripping out for Model I—which has no friction—while Fig. 4.6 does the same for Model II—which has friction. All displacements were taken from a midway point, located at $x = 500$ m.

As can be seen in Fig. 4.5, there is no difference between tripping in and tripping out when the model does not have friction. However, as seen in Fig. 4.6, when friction is added to the model the horizontal displacements become different. The horizontal displacement is greater during tripping in than tripping out. Since the column can buckle during tripping in, there will be a greater contraction and expansion in the horizontal direction due to the formation and dissipation of helices, thus causing greater amplitudes of displacement than on the tripping out case. Also, the displacements become more dependent on each other after the friction is added to the model; this explains why the horizontal displacement is almost unaffected by the angular

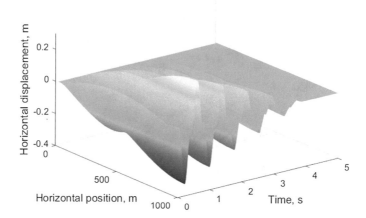

Fig. 4.4 Horizontal displacement, in meters, as a function of both horizontal position, in meters, and time, in seconds

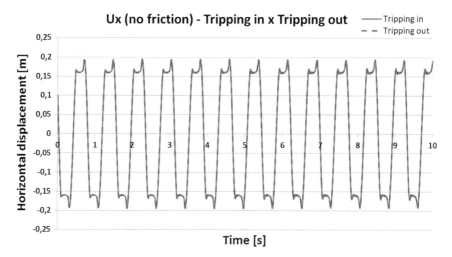

Fig. 4.5 Comparison of the horizontal displacement for tripping in and tripping out, in meters, as a function of time, in seconds, for Model I. The displacements used are for the point at x = 500 m

displacement in Fig. 4.5, in which there is no friction, but later is influenced by it in Fig. 4.6, in which there is friction. Finally, comparing Figs. 4.5 and 4.6, it can be seen that the motion is dissipated through time in Model II; this is coherent, considering that in Model I there are no dissipative forces while in Model II the friction force—which is dissipative—is acting.

Figure 4.7 shows a comparison of the angular displacement between Models I and II. Since the column can only suffer angular displacements during tripping in,

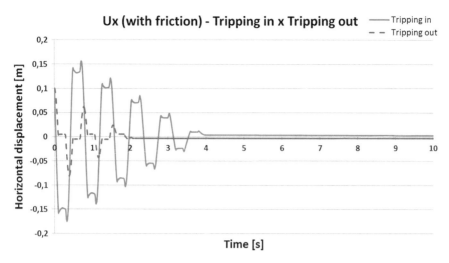

Fig. 4.6 Comparison of the horizontal displacement for tripping in and tripping out, in meters, as a function of time, in seconds, for Model II. The displacements used are for the point at $x = 500$ m

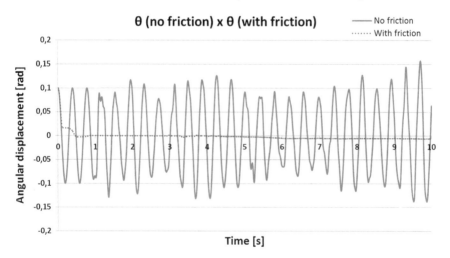

Fig. 4.7 Comparison of the angular displacement for Models I and II, in radians, as a function of time, in seconds. The displacements used are for the point at $x = 500$ m

there is no comparison to be done with the tripping out case. The data is taken again from a point located at $x = 500$ m.

As seen in Fig. 4.7, the angular displacement will not cease for Model I since there is no dissipative force, while for Model II the displacement will approach zero quickly due to friction. The angular displacements can be seen in Fig. 4.8 through another perspective, as a function of the position instead of time. The values were taken for $t = 1$ s.

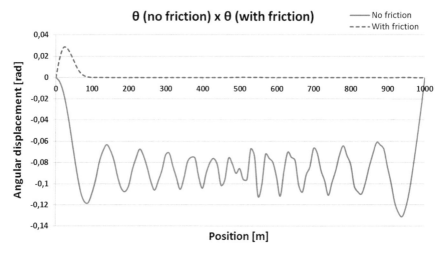

Fig. 4.8 Comparison of the angular displacement for Models I and II, in radians, as a function of position, in meters. The displacements used are for the time instant t = 1 s

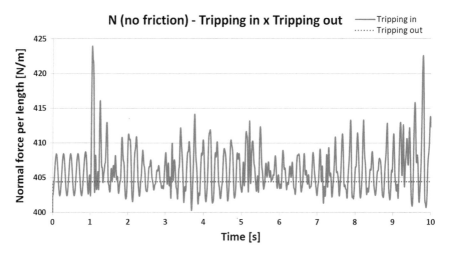

Fig. 4.9 Comparison of the normal contact force per unit of length for tripping in and tripping out, in Newton per meter, as a function of time, in seconds, for Model I. The forces used are for the point at x = 500 m

As seen once more in Fig. 4.8, the angular displacement is greatly dissipated due to friction. Even though the angular displacement becomes small, it is sufficient to cause a difference on the horizontal displacement, as observed in Fig. 4.6.

Figures 4.9 and 4.10 draw the same comparison as Figs. 4.5 and 4.7, but now for the normal contact force per unit of length. Once more, all the results were taken from a point located at x = 500 m.

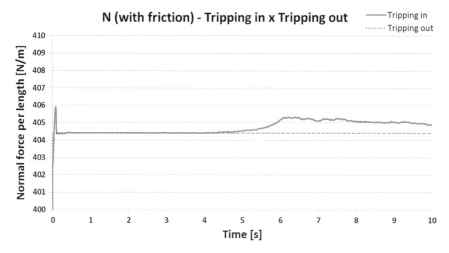

Fig. 4.10 Comparison of the normal contact force per unit of length for tripping in and tripping out, in Newton per meter, as a function of time, in seconds, for Model II. The forces used are for the point at $x = 500$ m

As seen in Fig. 4.9, the normal force fluctuates on tripping in, due to the angular displacement behavior seen in Figs. 4.7 and 4.8, in opposition to tripping out, where it remains almost constant. However, since the angular displacement is severely dissipated due to friction in Model II, the fluctuations disappear in Fig. 4.10. This can be easily explained by looking again at Figs. 4.7 and 4.8. Since the angular displacement is reduced close to zero for Model II, the tripping in case approaches the tripping out case, thus leading to what is seen in Fig. 4.10: the normal force for tripping in approaches the normal force for tripping out. It is still inconclusive if this small difference in normal forces can lead to a difference in friction forces; therefore, the friction forces must be compared as well.

Figure 4.11 provides a comparison between the friction forces for tripping in and tripping out in Model II. The friction force shown is the total force acting through the whole column. The graph is cut at $t = 2$ s—the time instant in which the transient part begins to vanish for tripping out—because there is a numerical instability after the dynamic solution converges to the static solution. This happens because the discretization used has low order; consequently, the errors pile up with the static solution.

As can be seen in Fig. 4.11, the friction forces are indeed different for tripping in and tripping out, thus being in agreement with the main hypothesis of this book. Looking again at Sect. 3.2, the friction force is dependent on the horizontal displacement, the angular displacement, and the normal contact force. Therefore, even though the angular displacement and the normal force for tripping in approach the tripping out case, the difference is enough to cause a significant difference on the horizontal displacement, which then leads to a difference on the friction forces. Comparing Eqs. 3.241 and 3.235 through their $\hat{i}$ component, it can be seen that the difference between

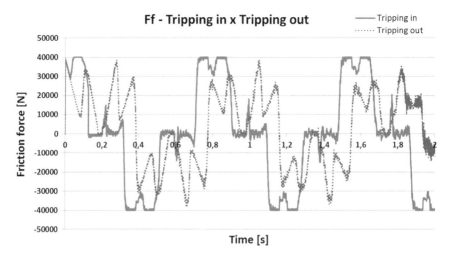

Fig. 4.11 Comparison of the friction force for tripping in and tripping out, in Newton, as a function of time, in seconds, for Model II. The values shown are the total force acting through the whole column

tripping in and out is given by the term $C_f = \left|\frac{\partial u_x}{\partial t}\right| / \sqrt{\left(\frac{\partial u_x}{\partial t}\right)^2 + r^2\left(\frac{\partial \theta}{\partial t}\right)^2}$. As long as $C_f \neq 1$, the friction forces will be different; to have $C_f \neq 1$ means that the angular displacement must be $\theta \neq 0$. This conclusion aligns with the hypothesis used previously—that the ratio of the friction coefficients on the axial and lateral directions is proportional to the ratio of the velocities on said directions. Therefore, since the normal contact force in both cases is not different—as seen in Fig. 4.10—the real cause for the different friction forces is the angular displacement—and, indirectly, the axial displacement. Another comparison of the friction forces can be seen in Fig. 4.12, now by taking the force per unit of length as a function of the position. The data is again taken from $t = 1$ s.

As can be seen in Fig. 4.12, while on tripping out the friction force per unit of length remains almost constant, it fluctuates heavily for the tripping in case, thus causing a difference on the total friction force when the contributions at each point are added up. This is another aspect contributing toward the result shown in Fig. 4.11.

For Models III and IV, the horizontal well trajectory shown in Fig. 4.2 will be used. Starting with Model III, Figs. 4.13 and 4.14 show the horizontal displacement for several inclinations for the tripping in case and the tripping out case, respectively. The inclinations were taken from specific points along the well trajectory: $\alpha = 0°$ at $x = 500$ m on the vertical segment, $\alpha = 37°$ at $x = 1500$ m on the first build-up segment, $\alpha = 55°$ at $x = 2000$ m on the slant segment, $\alpha = 73°$ at $x = 2500$ m on the second build-up segment and $\alpha = 90°$ at $x = 3000$ m on the horizontal segment.

As can be observed from both Figs. 4.13 and 4.14, the displacements are greater the higher the angle, remembering that $0°$ is the vertical position and $90°$ is the horizontal position. This provides an interesting result when coupled with Fig. 4.15

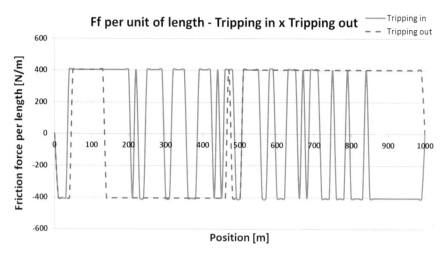

Fig. 4.12 Comparison of the friction force for tripping in and tripping out, in Newton per meter, as a function of position, in meters, for Model II. The forces used are for the time instant t = 1 s

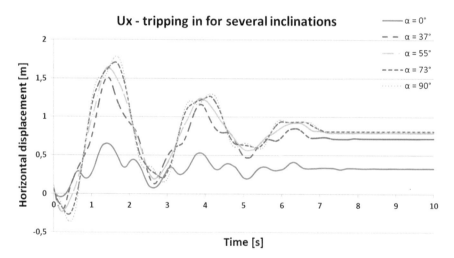

Fig. 4.13 Comparison of the horizontal displacement during tripping in for several inclinations, in meters, as a function of time, in seconds, for Model III. The data is taken from points located at x = 500 m, x = 1500 m, x = 2000 m, x = 2500 m, and x = 3000 m

below. Figure 4.15 shows the normal contact force per unit of length during tripping in for several inclinations. It can be seen that the greater the angle, the higher the normal force; this makes sense physically considering that the contact weakens as the column moves to the vertical position, thus reducing the normal force. If the normal force reduces, so does the friction force, which is directly proportional; finally, if the friction forces reduce as the column approaches the vertical position, the horizontal displacement should be higher for small angles, but this is not the case as observed

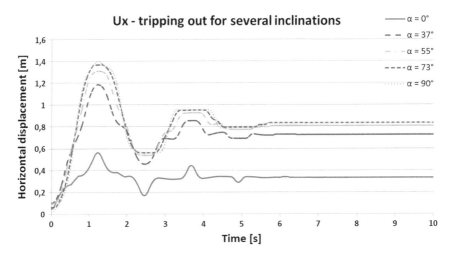

Fig. 4.14 Comparison of the horizontal displacement during tripping out for several inclinations, in meters, as a function of time, in seconds, for Model III. The data is taken from points located at x = 500 m, x = 1500 m, x = 2000 m, x = 2500 m, and x = 3000 m

in Figs. 4.13 and 4.14. Instead, the opposite occurs: the points with greater angles, which are located deep down on the well, are the ones with greater displacements. To check the aforementioned result, the same column length was simulated for five scenarios with constant inclination using the five angles from Figs. 4.13 and 4.14, and then the time histories of the same five points at the same five depths were taken—one point from each of the five scenarios. This is shown in Fig. 4.16 below. Except for α = 0°, in this case, the expected outcome was observed: the higher displacements are at lower angles, in which the friction force is smaller, and thus it dissipates less. It was also observed that for small inclinations, greater displacements are seen when a segment with constant inclination is used rather than when the full trajectory is applied; meanwhile, for large inclinations, greater displacements are seen when the full trajectory is used instead. This comparison between the two analyses shows that the well curvature combined with the buckling effect is affecting the axial behavior of the column by diminishing displacements for small angles while causing greater displacements for larger angles. The friction being distributed throughout the whole well can explain this: the friction at small angles is actually higher than initially thought, causing smaller displacements; meanwhile, the friction at large angles is actually lesser than initially thought, causing larger displacements. Therefore, this interesting result can only be seen when a full trajectory is applied to the model and would not be seen otherwise if a segment with constant inclination was simulated instead. In addition, as in Model II, the displacements dissipate due to friction and remain on a stationary value, which is the static displacement of the column.

Figure 4.17 shows the angular displacement during tripping in for several inclinations. Except for α = 0°, the results are in good agreement with Figs. 4.7 and 4.8 for Models I and II. For α = 0°, the angular displacement possesses an unusual behavior,

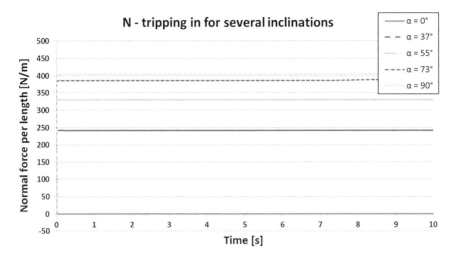

Fig. 4.15 Comparison of the normal contact force per unit of length during tripping in for several inclinations, in Newton per meter, as a function of time, in seconds, for Model III. The data is taken from points located at x = 500 m, x = 1500 m, x = 2000 m, x = 2500 m, and x = 3000 m

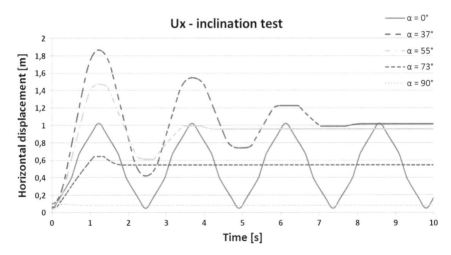

Fig. 4.16 Axial displacement for different inclinations for comparison with Figs. 4.13 and 4.14, in meters, as a function of time, in seconds, for Model III. The data is taken from points located at x = 500 m, x = 1500 m, x = 2000 m, x = 2500 m, and x = 3000 m

which is to be expected considering the hypotheses of this model. As mentioned before, the column is considered to always remain in contact with the wellbore, but this may not be entirely true for the whole vertical segment where α = 0°, despite being a good hypothesis for almost the whole range between 0° and 90°. Since the angular displacement caused by this effect is not completely out-of-scale when compared with the angular displacements at other points, the results remain valid.

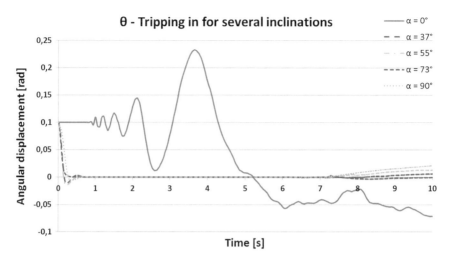

Fig. 4.17 Comparison of the angular displacement during tripping in for several inclinations, in radians, as a function of time, in seconds, for Model III. The data is taken from points located at $x = 500$ m, $x = 1500$ m, $x = 2000$ m, $x = 2500$ m, and $x = 3000$ m

For a deeper analysis regarding the angle, a more refined graph can be made. Figure 4.18 shows the angular displacement for small angle values: $\alpha = 0°$, $\alpha = 2°$, $\alpha = 5°$, and $\alpha = 10°$. The effect observed in Fig. 4.17 is more pronounced for angles up to 2°; at 5°, the angular displacement dissipates more quickly, while at 10° the expect behavior from Fig. 4.17 already occurs. Therefore, the solution seems to be in accordance for $\alpha > 2°$, while for $\alpha < 2°$ the solution can still be used but loses precision due to the hypothesis used.

Finally, Fig. 4.19 shows the total friction force comparison during tripping in and tripping out for Model III by adding up the contributions through the whole well trajectory. Once more, the total friction force remains different for both tripping in and tripping out cases, thus further agreeing with the initial hypothesis of this book. Again, the graph was cut—now at $t = 5$ s—for the same reason as Fig. 4.11.

Now moving to Model IV, similar results are achieved as can be seen in Figs. 4.20 and 4.21 for the horizontal displacements, during tripping in and tripping out for several inclinations. The only difference regarding Models III and IV is that in Model IV, the variables no longer reach a stationary value, due to the presence of a periodic excitation—the heave motion. The solution can be divided now into two parts: the transient solution, which contains the system characteristics, and the permanent solution, which contains the external periodic excitation characteristics. As can be seen in Figs. 4.20 and 4.21, a periodic motion starts around 6 s, which is a consequence of the heave motion. For a heave amplitude of 0.5 m, it can be seen that the heave motion becomes significant on the overall response of the system.

Once more, the total friction force comparison during tripping in and tripping out is shown in Fig. 4.22, now for Model IV. The total friction force is still different for

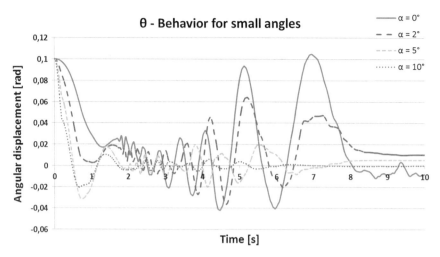

Fig. 4.18 Angular displacement θ for small angles, in radians, as a function of time, in seconds, for Model III. The data is taken from points located at $x = 950$ m, $x = 980$ m, $x = 1030$ m, and $x = 1100$ m

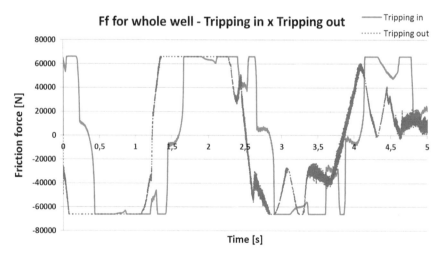

Fig. 4.19 Comparison of the total friction force for the whole well trajectory during tripping in and tripping out, in Newton, as a function of time, in seconds, for Model III

both tripping in and tripping out cases, thus still agreeing with the initial hypothesis of this book.

Last, a quick comparison between the horizontal displacement of Models III and IV is made in Fig. 4.23, using the displacements for $\alpha = 37°$. As already explained, the only difference between the two models is the presence of a periodic motion on Model IV response once the transient part of the solution vanishes—around 6 s.

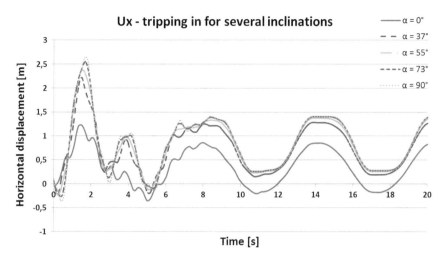

Fig. 4.20 Comparison of the horizontal displacement during tripping in for several inclinations, in meters, as a function of time, in seconds, for Model IV. The data is taken from points located at x = 500 m, x = 1500 m, x = 2000 m, x = 2500 m, and x = 3000 m

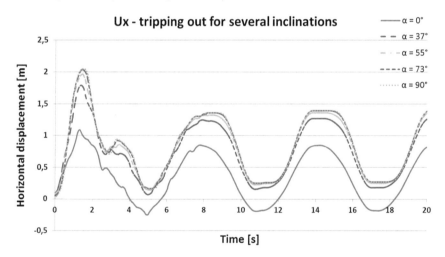

Fig. 4.21 Comparison of the horizontal displacement during tripping out for several inclinations, in meters, as a function of time, in seconds, for Model IV. The data is taken from points located at x = 500 m, x = 1500 m, x = 2000 m, x = 2500 m, and x = 3000 m

4.3   Final Remarks

In this book, we developed a dynamic model to understand the behavior of columns constrained inside directional wells during completion operations, such as running a tubing inside a cased hole, running a coiled tubing inside a tubing or running a sand screen inside an open hole. The development was done through four different

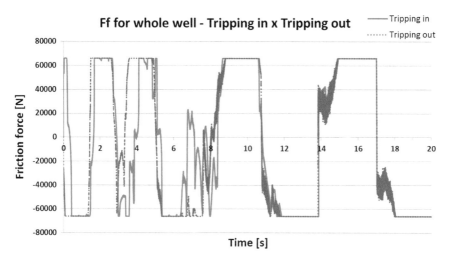

Fig. 4.22 Comparison of the total friction force for the whole well trajectory during tripping in and tripping out, in Newton, as a function of time, in seconds, for Model IV

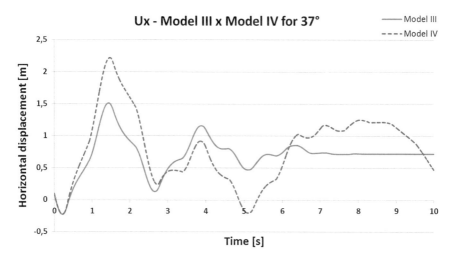

Fig. 4.23 Comparison of the horizontal displacement, in meters, as a function of time, in seconds, between Models III and IV. The data is taken from a point located at x = 1500 m

models, with each one increasing the problem complexity: Model I considered a frictionless column inside a horizontal portion of well; Model II added the effect of friction to the problem; Model III considered the well inclination, thus being able to analyze the behavior of a column inside any well trajectory; and, finally, Model IV considered the effect of heave motion transmitted through the column, thus moving from an onshore to an offshore environment.

   The results for Models I and II, given by Figs. 4.3 through 4.12, show that, in fact, the friction force is different during tripping in and tripping out a column inside a well. Therefore, the results are in good agreement with this book initial hypothesis that the difference on the friction forces during tripping in and out is a consequence of the dynamic buckling of the column. In addition, the effect of the friction force on the variables was shown: the friction dissipates the motion, thus turning a solution that was initially permanent into a transient one, with the response decaying to its static response as time passes.

   The effect of the well inclination—Model III—was seen also on Figs. 4.13 through 4.19. An interesting effect was observed on Figs. 4.13 through 4.16: despite the normal contact force decreasing as the well inclination becomes closer to vertical—which reduces the friction force and thus would increase the horizontal displacement—a decrease on the horizontal displacement was seen instead. This is a result that can be seen only if the full well trajectory is applied to the model, in opposition to simulating separately several well segments of constant inclination. Also, the hypothesis of the column being in contact with the wellbore through its whole length is not entirely valid for angles too close to 0°, as seen on Figs. 4.17 and 4.18.

   Last, the effect of the heave motion—Model IV—is seen on Figs. 4.20 through 4.23. As discussed, the only effect that the heave causes is introducing a permanent component on the solution of this system. Instead of dissipating its motion entirely, the column remains vibrating indefinitely thanks to the heave motion, with the same angular frequency of the heave and with amplitude that varies along the column length.

   For future works, a validation with real data is still needed. Currently, data collected specifically in order to solve this problem is still taken with a static mindset—which means that no variable involved is measured as a function of time, only as a function of position. With these models, we intend to raise awareness on the issue of dynamic buckling while also hoping to acquire real data for validation in the future.

   Also, we suggest improving the hypotheses presented through the models, such as including the effect of external and internal fluid, since they will induce both viscous damping—which attenuates the column vibration and consequently the buckling effect—and a buoyancy force; considering that the column no longer needs to remain in contact with the wellbore, which is necessary not only for accurately describing vertical segments of well but also for inclined segments, since the column may lose contact along the trajectory; considering that the column is actually moving forward or backward—and not only vibrating around an equilibrium position—while being assembled or disassembled, respectively; finally, improving the finite difference discretization or even propose a finite elements discretization. In this book, the discretization used was of the simplest form available with the lowest order; more complex and higher-order discretization such as Runge–Kutta can be employed to obtain results that are more accurate.

Last, we believe that these models can be applied to other problems—in which a column is constrained inside another column—as long as some adaptions are made, such as in the boundary conditions. For example, this model could be used to describe the dynamics of the sucker-rod pumping method, which is used for artificially lifting from a well.

4.4 Summary

This chapter:

- Presents a methodology for analyzing a case study;
- Presents a case study, with results and discussions;
- Presents final remarks and proposes topics in order to push forward the subject.